"The story of how a seemingly small innovation—the like button— was birthed, surged into prominence, and went on to change whole industries. Innovation in technology, Reeves and Goodson tell us, is by no means orderly, planned, and predictable. Rather it is disorderly, unruly, and unguided. A riveting tale."

—**W. BRIAN ARTHUR**, author, *The Nature of Technology*

"Moving from ancient Roman gestures to modern-day social media empires, Reeves and Goodson brilliantly trace how the like button became the foundation of our digital experience. It's an essential read for anyone interested in understanding how seemingly minor innovations can reshape not just business but human interaction."

—**AZEEM AZHAR**, founder, *Exponential View*; Digital Fellow, Stanford University

"Like, WOW!"

—**PAUL BUCHHEIT**, creator, Gmail; cofounder, FriendFeed

"Two thumbs up! A must-read for those interested in how our individual digital interactions shape our world view."

—**DAVID DE ROTHSCHILD**, British adventurer and environmentalist; Chairman and Trustee, Voice for Nature Foundation

"*Like* is a captivating exploration of the like button's evolution, revealing how this humble icon transformed the way people all over the world interact. A must-read for those intrigued by the power of simple innovations to influence society."

—**KARIM R. LAKHANI**, Dorothy and Michael Hintze Professor of Business Administration, Harvard Business School; coauthor, *Competing in the Age of AI*

"Through the fascinating story of the like button, the authors show us the bigger picture of how tech innovation works, how industries are created and disrupted, and how unintended consequences arise and can be dealt with through adaptive regulation."

—**KAI-FU LEE**, coauthor, *AI 2041*

"*Like* is a must-read for anyone interested in how digital technology is shaping our lives and re-wiring our brains. Fascinating, fast-paced, and full of original insights, this book will open your eyes to a much richer and more textured story of the rise of social media, including where it began, how it works, and what tomorrow may bring. Two thumbs up!"

—**ANNA LEMBKE**, MD, *New York Times* bestselling author, *Dopamine Nation*

"*Like* is wildly original. Who would have thought that such a simple button could hide a deep understanding of technology, money, and regulation? Martin Reeves shows once again that he is a unique thinker, possessing one of the most valuable human qualities: curiosity."

—**NIELS LUNDE**, Editor in Chief, Dagbladet Børsen

"If you're looking for a book to 'like' I strongly recommend this one. It's a can't-put-it-down page-turner about the significance of something (and its serendipitous route into the world) you probably never appreciated. It's full of twists, turns, interesting characters, and even biology and neuroscience. A fun and instructive read!"

—**RITA McGRATH**, strategist; professor, Columbia Business School; and author, *Seeing Around Corners* and *The End of Competitive Advantage*

"Don't press the like button! Buy this book instead."

—**GARY SHTEYNGART**, bestselling author, *Super Sad True Love Story*

"This book isn't just about a button—it's a compelling journey through the messy process of innovation and a must-read for anyone curious about the hidden stories behind the things we click every day."

—**BIZ STONE**, cofounder, Twitter

"Terrific, fun, and keenly illuminating. Reeves and Goodson tell the remarkable tale of the like button—and use it as the foundation for a host of insights into innovation, the brain, and our species."

—**CASS R. SUNSTEIN**, author, *How to Become Famous*; coauthor, *Nudge*

"Reeves and Goodson trace the curious origins of the like button and how it facilitated the creation of a new, multibillion dollar industry. The connections between innovation, technology, anthropology, brain science, and cultural history are fascinating. A must-read!"

—**TIGER TYAGARAJAN**, former President and CEO, Genpact

LIKE

THE BUTTON THAT CHANGED THE WORLD

MARTIN
REEVES

BOB
GOODSON

LIKE

HARVARD BUSINESS REVIEW PRESS · BOSTON, MASSACHUSETTS

The web addresses referenced in this book were live and correct at the time of the book's publication but may be subject to change.

Library of Congress Cataloging-in-Publication Data

Names: Reeves, Martin author | Goodson, Robert, 1980- author
Title: Like : the button that changed the world / Martin Reeves and Robert Goodson.
Description: Boston, Massachusetts : Harvard Business Review Press, [2025] | Includes index. |
Identifiers: LCCN 2024045041 (print) | LCCN 2024045042 (ebook) | ISBN 9798892790451 hardcover | ISBN 9798892790468 epub
Subjects: LCSH: Internet marketing—Psychological aspects | Social media—Psychological aspects | Social media—Economic aspects | Social media and society | Praise in social media | Business—Technological innovations
Classification: LCC HF5415.1265 .R446 2025 (print) | LCC HF5415.1265 (ebook) | DDC 302.23/1019—dc23/eng/20250106
LC record available at https://lccn.loc.gov/2024045041
LC ebook record available at https://lccn.loc.gov/2024045042

ISBN: 979-8-89279-045-1
eISBN: 979-8-89279-046-8

For all the tinkerers and makers
that history will never record.

Contents

Chapter 3

Why the Thumb?   75

A button for registering approval could have taken many
forms. Why had early innovators' minds gone in the direction
of a thumbs-up, and why did others gravitate to that same
symbol? This chapter tells the tale.

Chapter 4

Why Do We Like Likes?   93

What was really going on in the explosive adoption of the like?
This chapter draws on evolutionary biology, anthropology,
social psychology, neuroscience, and other human-centered
disciplines to examine just why we like likes.

Chapter 5

What Happens When You Click?   109

How does the machinery of liking actually work? What happens when you hit a like button? Take a peek inside the code
to find out.

Chapter 6

The Business of Likes   131

The people who are getting gratification from the button—the
likers and the liked—generally aren't paying a cent for that
pleasure. So who *is* paying for it all, and why? This chapter
digs deep into the business of likes: the various models of how
liking generates revenues and wealth.

Chapter 7

Unintended Consequences 165

The use of social media comes with potential negative impacts on privacy, mental health, and social cohesion, and the like button was pivotal in the meteoric rise of this new industry. In this chapter, we examine how such unintended outcomes arise and how we can better manage them.

Chapter 8

The Future of Likes 195

How will liking play out in next-generation digital social environments? Will new norms take hold for where and how to register a like? The book closes with some speculation and early evidence about how liking will happen in the future.

LIKE

Why a Book about the Like Button?

Have you ever paused to wonder how the digital thumbs-up icon you have likely clicked multiple times today was invented or how it works? Probably not, we guess. Indeed, why would you waste a moment thinking about this tiny piece of technology?

The like button seemed equally unremarkable to us—until it serendipitously cropped up in a conversation. Bob Goodson (a Silicon Valley veteran educated as a classical literature scholar) and Martin Reeves (a business strategy adviser, researcher, and author) had recently become friends, Bob was moving house, and he happened to mention an old sketch he had stumbled across while packing (figure I-1). It depicted a prototype for the like button, made when Bob was one of the first employees at Yelp. Intriguingly, the date on the

FIGURE I-1

Bob Goodson's original sketch of the like button

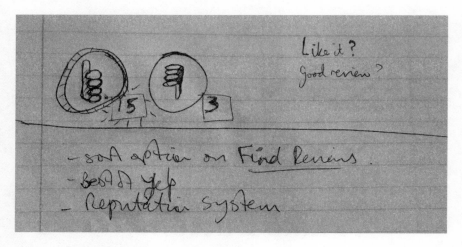

Source: Photo by author (Bob Goodson), courtesy of Yelp

sketch (May 18, 2005) preceded the adoption of the like button by Facebook by several years.

This triggered a daylong conversation and a subsequent three-year research journey. Not only did we dive into the origins and mechanics of the like button and the pivotal role it played in the rise of social media and digital marketing, but we also wanted to examine what the tale of the like button teaches us about how technological innovation and regulation work in practice. As the answers to one question generated the next, our research led to conversations with experts in numerous fields of knowledge, including technology and business history, the cultural and anthropological history of the thumbs-up gesture, the evolutionary neuroscience of sociality, the sociology and mathematics of innovation, and the regulation of new technologies.

Our journey leaves us believing that the humble like button is in fact worthy of your attention, and this book explores why. The like button was pivotal in the rise of the $220 billion social media industry and the digital transformation of advertising and marketing. It also tells us a lot about ourselves—the origins and consequences of our deeply social nature and how we use gesture and language accordingly. It reveals how the messy process of innovation works in practice. It illuminates why there are always unintended consequences of technological innovation and how we can go about managing them. And it provides some hints on what might come next in digital sociality and business. We share the story in roughly the sequence it unfolded for us, resisting the temptation to summarize the lessons in a tidy framework and instead preserving the messy but edifying connections between biology, behavior, and business.

In chapter 1, we delve into the invention of the like button. Spoiler alert: you may know it best from Facebook, but that site's adoption of the feature only happened well into the scale-up phase of the like button's life, when it was already obvious it was the social media equivalent of catnip.

In chapter 2, you'll read how our initial focus on the like button turned into a broader reflection on how stuff gets conceived and built in Silicon Valley and other innovation hubs. Having learned that Bob was involved in the genesis of this truly iconic feature, Martin was at first amazed that Bob himself appeared only barely cognizant of that fact. Yes, it's always difficult to reconstruct the past, but even at the time, as a central actor, Bob apparently hadn't been consciously thinking "This could be big" or noticing how different pockets of work were already overlapping and colliding to produce something new and important. He was just a guy in a startup, puzzling over that week's interface design challenge. Having since interviewed many of the key people involved in the like button's origin, we found they all shared this reaction and perspective. Looking at the like button from this angle, we could see it as a perfect case study of the dispersed, disorderly, serendipitous, unguided process from which high-impact innovations usually emerge—despite how such stories are often made to sound in the retelling.

One thing that intrigued us at the outset was that a button for registering approval could have taken a lot of forms. Clapping hands, for example, or a smile, or a heart, or maybe a gold star. So why the thumbs-up? It turns out there are all kinds of interesting reasons this gestural symbol is a deep-seated meme in human culture. And surprisingly, the one thing from antiquity people think popularized the gesture—the emperor presiding over gladiatorial contests in the Roman Coliseum—probably didn't involve an upturned thumb at all. Chapter 3 tells the tale.

Part of why the like button took off so fast was that the icon itself resonated. But what was really going on in the explosion of adoption that followed? What were users of social sites getting out of a like that made it so satisfying and even addicting? This is the question taken up in chapter 4, which draws on evolutionary biology, anthropology, and social psychology to examine why we like liking.

Chapter 5 considers what happens when users hit the like button and what happens when they see that something they've posted was liked. How does the whole infrastructure of liking work? When you register an emotion on a digital platform, what does that activate in the software—where does your like go next in the machinery of social media? And what else gets triggered down the line?

Today, people are collectively hitting like buttons billions of times per day, and all those likes and the machinery processing them obviously add up to a lot of bits and bytes zooming around a large ecosystem of partners. A massive investment in hardware and software systems enables this. Yet the people who are getting gratification from the button—the likers and the liked—generally aren't paying a cent for that pleasure. So who *is* paying for it all, and why? What's their hope of getting a return on their investment in massive computer arrays and communication networks—and is that hope being fulfilled? How exactly, and in how many ways, are friendly likes being transformed into monetary gain? Chapter 6 digs deep into the business of likes and various ways in which liking generates revenues and wealth.

Critics of social media's rapid incursion into our lives point to its presumed negative impacts on privacy, mental health, social cohesion, and even democracy. How did such unforeseen consequences arise after the like button was unleashed on the world? How did we go from a useful little symbol on a website to the perversion of democratic campaigns? Net-net, are we better off because we have it in our lives? Chapter 7 explores the timeless problem of the unintended effects of all technology innovations and examines what the like button can teach us about how such outcomes arise and how we can better manage them.

We close this book with speculations and some early evidence of how liking will play out in next-generation digital social environments. As people gravitate to new media and platforms with ever-greater richness and dimensionality, will their emotive reactions still be communicated in ways so easy to tally? If today's social media paradigm of intermittent, self-contained posts transforms into some more fluid and boundless model, will new norms take hold for where and how to register a like? As we witness the rise of generative AI, what role will the data stored from all those likes have to play? As the prospect of brain-computer interfaces becomes more feasible, could a like be registered from thought? Chapter 8 speculates about the plausible futures of the like button.

Our hope is that by following the winding journey of the like button from our early evolutionary history to the future of digital society, you will gain some of the same appreciation of the deeply human underpinnings of technology and the messy process of innovation that we did.

Chapter 1

The Invention of the Like Button

As many times as you've pressed a like button in the past couple of decades, has it occurred to you to wonder who came up with it—or why or when or how? Probably not, and frankly it had never occurred to either of us prior to the wide-ranging conversation we had in 2022. More surprisingly, it also wasn't a piece of history Bob himself had spent much time pondering—despite his being very much part of the story. When we realized the answers weren't so easy to pinpoint, the questions only got more intriguing. As it turns out, there was a whole cast of *whos*, a variety of *whys*, and plenty of points along the timeline competing for the definitive *when*.

But maybe we should start with the *what* of the matter. The definition of the term *like button* might seem as obvious as the thumb symbol on your screen, but some higher-resolution clarity can only help: it's a simple one-click mechanism that allows a user of a site or platform to register an emotional reaction to a piece of content created by another user. It also lets the creator of the content, and often everyone else, see that reaction, and it keeps track of how many people have liked it altogether. Note that this isn't the same as a product rating or seller rating on an online retailer's site. It's specific to the social web. And in that context, it's not the same as friending, following, or mentioning a person or the same as reposting a piece of content. All these functions might be offered on a platform, but they address distinct needs. The like button just says "I like this."

But that tiny little task allowed the like button to get very big—and get big very fast. Accounts of how it all happened vary in people's memories and in the few media stories that have been published. But Bob had a front-row seat because, back in 2003, before any such feature had been integrated into any social media user interface, he was working as employee number one at a dawn-of-the-social-web startup called Yelp. Why, exactly, was Bob, as an Oxford graduate student in medieval studies, even in that unlikely position? The answer starts with the mind of Ukrainian American software engineer and businessman Max Levchin.

· · ·

Max Levchin is a big-deal internet business builder in Silicon Valley. Born in Kyiv, Ukraine, he came to America as a sixteen-year-old. By age twenty-three he had cofounded, along with Peter Thiel and Elon Musk, the company that turned into PayPal. After that company was sold to eBay for $1.5 billion in 2002, Levchin stayed on for some months to help with eBay's integration of the acquisition. When it was time to move on, he was still a young man but now a very wealthy one and was thinking hard about what new businesses he could build next.

For his part, Bob was a computer enthusiast—he'd been programming video games since age eight and starting companies since he was seventeen, but he had lately been elbow deep in illuminated vellum manuscripts. As an undergraduate, he had studied philosophy, then he had gone on to Oxford University for graduate studies in medieval literature. His thesis research focused on fifteenth-century cento—an obscure form of poetry in which the author constructs new verses out of lines written by previous poets. It wasn't an obvious educational path toward a career in an entrepreneurial internet business.

But this was a very special time in Silicon Valley history because there weren't yet obvious training grounds for what was to come next. It was a moment when a kind of wild mixing of disciplines was happening. And Max Levchin, having exited PayPal, had just come to the United Kingdom at the invitation of a good friend and fellow tech entrepreneur. Investors who had been part of PayPal, and others who wished they had been, were constantly calling Max to ask what he was going to do next. His friend James thought the trip would help Max clear his mind and get focused on a new plan. As part of his itinerary, he would give a talk to a student startup club that Bob had recently helped found, called Oxford Entrepreneurs.

Levchin had no shortage of ideas. As Bob sat listening in that lecture hall, he found it hard not to get excited about the next wave of internet opportunities Levchin was describing. Today, in retrospect, it's clear that Max foresaw how big the shift would be to consumer-oriented social web development, the so-called Web 2.0 era, and his timing was brilliant.[1] After the talk, Bob approached Levchin, and the two fell quickly into a conversation about social apps people might love to use. That initial discussion turned into follow-up correspondence. Now, at his new friend's invitation, Bob had come to Silicon Valley for the first time to soak up some of the spirit of the place. He'd been tagging along as Levchin went from meeting to meeting, having all kinds of interesting conversations.

A few days into that visit, the two sat down at a coffee shop to compare notes. As Bob recalls, one thing really struck him about Max as they started talking was that his head was surrounded by a massive doughnut. The huge plastic doughnut was hanging from the ceiling behind him, but it was exactly framing him in a way that was kind of cosmic and mesmerizing. The scene turned even more surreal a few minutes into the conversation, when something also dawned on Bob as he responded to questions Max was throwing at him: Bob was evidently being considered for some kind of job.

Max was none too clear about what he thought Bob might stick around and do, but he was sharing his thoughts about how high-impact innovations could emerge from startup incubators. He had just decided to launch one himself. At his MRL Ventures, the idea was to hatch new ventures by funding entrepreneurial people even before they had settled on great product ideas. His model was that the incubator would identify broad areas of opportunity, then bring together some capable, creative people and pay them a salary while they pulled together specific ideas in those areas and have teams go after them. Then, if a team could pitch what sounded like a potential winner, MRL would put up the first million dollars to get it off the ground.

In some ways it was a terrible time to be trying this kind of approach. Not much time had passed since the dot-bomb crash of 1999–2000, when billions in market capitalization had evaporated into thin air and most investors with funds to spare were playing it safe. Promising talent wasn't easy to come by, either. There were plenty of people with valuable experience, but most had landed at larger companies and were holding tight. The nuclear winter that had descended on the web had yet to see a thaw. Max Levchin was actually excited about the bleakness and pessimism all around him, seeing the opportunity in a moment that would not last. Friendster had just been launched in 2002, and its rapid uptake was showing the promise of social media. Still, he was finding it hard to get people to leave stable positions for something as uncertain as a stint at an incubator.

This was the situation, then, in which a medievalist could arrive in Silicon Valley and appeal to a renowned internet business founder. But there was another reason that Max wanted Bob to join his merry band. It was a new era of internet business building, and Bob had no baggage—no preconceived model of how a Silicon Valley startup worked or should work. By that point, Max had already mapped out the broad territories MRL Ventures should address. Given his expansive vision of the coming social web, he knew there would be more opportunities than one group could possibly have the bandwidth to pursue, so the important thing was to stake some early claims in those that could scale the fastest and biggest. This approach narrowed the goal to building high-impact businesses in four realms:

- **Build an advertising business designed for this emerging environment.** This brilliant objective in 2003 anticipated the explosion that would soon follow in online marketing. (But perhaps it was ahead of its time: the business MRL incubated into AdRoll didn't attain the multibillion-dollar level of success that others later would.)

- **Provide a way for people to chat online.** That effort, led by Alec Matusis, would soon produce Chatango, a plug-in that anyone could employ to let visitors to their page get into a conversation about what they saw there. Again, this was prescient—Twitter would not appear for another two years— and it involved a technical breakthrough, using Macromedia Flash (normally used to create animated elements) for a purpose no one had previously recognized as possible. But it would soon fall victim to its own success: when Chatango suddenly went viral among teenagers, eBay (which had been the biggest driver of its uptake) suddenly banned the application, and it never fully recovered from the blow, although it is still running today.

- **Invest in image sharing.** Levchin knew that image sharing would see explosive growth on the social web as bandwidth increased. Here, MRL Ventures was responsible for Slide, which Max himself led the charge on and launched in 2004. It was a hit in the early days of Myspace and Facebook before being sold in 2010 (for hundreds of millions of dollars) to Google.

- **Transform local search.** This effort would give consumers a better way than the traditional Yellow Pages to find the best small businesses and restaurants, health-care professionals, and other service providers in their locales. This was the genesis of Yelp: eyeing the $30 billion being spent per year on print Yellow Pages ads and imagining all the ways an online platform could improve the experience.

In 2003, Max had figured out areas of opportunity to target but hadn't figured out the products at all, let alone the business models. To create those, he wanted to bring together a community of problem-solvers who could build companies around solutions that would be valuable to people. So far, he had pulled in David Galbraith (who had just, as part of his last venture, coauthored the internet standard RSS 1.0) and Jared Kopf (another builder of various world's-first technologies in startups dating back to 1999) as experienced entrepreneurs in residence. Joining them were two other PayPal veterans, Jeremy Stoppelman and Russel Simmons. Stoppelman had been vice president of engineering at PayPal and had just enrolled in Harvard Business School when Max coaxed him back to California. Simmons had been PayPal's lead software architect.

Like these others, Bob got what Max was saying about the dawning consumer internet era and the unique opportunity the moment was presenting. He even took care to write down Max's words: "There is going to be a resurgence of the web, and this time around we can understand the business models. Broadband will change things fundamentally. And we're going to execute properly."[2] Almost before he knew it, Bob was part of the team getting to work on the local-search opportunity, as Stoppelman and Simmons's first employee and product manager.

. . .

Yelp is now a household name: In 2023, every month, the site was visited by some 128 million unique users who accessed its content via desktop, mobile, and app versions. It has nearly five thousand employees, earns over $1 billion in annual revenues, and, in late 2024, boasts a market capitalization north of $2 billion. It performs a valuable service that wasn't available before its appearance—and the biggest reason it was able to grow fast and get big was its early success in figuring out what would become known as user-generated content.

If you've used the site, you know it allows you to search for a type of business or organization in a geographical area. In response, Yelp delivers not only a list of results in that category but also any reviews, recommendations, and comments that have been posted by past customers of those places, with the most relevant posts appearing at the top. By 2023, Yelp's pages contained more than 265 million reviews—but at the beginning, it wasn't at all clear that numbers like this were possible.

In fact, getting Joe or Jane Public to write public reviews wasn't even the plan at the outset. When Stoppelman and Simmons accepted their first million dollars in seed money, the concept they had successfully pitched was for a platform on which friends would share recommendations with each other online, essentially taking the basic mechanism of word-of-mouth influence and giving it more structure and reach by digitizing it. When the site first launched in October 2004, this was the value proposition: Whenever you were trying to make a decision about where to spend your money—say, a restaurant to try next, for example, or a plumber to call for your leaky pipe—you could "send a Yelp" to your circle of friends, people who shared your tastes and had no incentive to steer you wrong. All your friends would then get a message along the lines of "Jill would like a recommendation for breakfast tacos in San Antonio" and be provided with access to a database where they could select a place they thought Jill would go for. Collectively, people's recommendations to each other would then add up to ratings for the sellers.

Unfortunately, it didn't work. Yelp launched, and it got some early use because of press buzz, but within a few months, it was evident it wouldn't gain traction. The central idea to generate reviews by emailing friends was found not to be viral. One big problem was the delay in the responses: by the time friends saw the emailed query, most figured you had already thrown up your hands and made your own choice. But more fundamentally, the problem was that Yelp's success rested on a three-legged stool of motivations: shoppers would have to want recommendations, businesses would have to care about their ratings, and people with knowledge of businesses would have to want to offer recommendations. And while the first two could be counted on, that third leg was coming up very short. What natural motivation did ordinary people have to act like critics and provide the input that would add up to ratings?

It was clear to Stoppelman and Simmons that, to make Yelp work, the central design challenge was to create a system that would generate a large and constantly refreshed reservoir of trustworthy reviews readily available for any searcher the moment they needed input to a decision. As that problem was the obvious nut to crack, and as the first approach had not worked, the team spent day and night debating alternative setups.

Yelp had stayed in the incubator as it started the first versions of its offering. The team spent its hours together in a small, narrow red brick office on Mission Street in San Francisco, nestled among towering skyscrapers housing the likes of Salesforce, PwC, and other big companies. This tiny building had somehow escaped a century's worth of earthquakes and, along the way, had been retrofitted with red metal beams that crisscrossed its entire interior. Overhead lights ran alongside exposed heating ducts and electric cables. Bob recalls the thick tape wrapped around the front of his desk, bright yellow with the word *trash* printed in black at regular intervals. All the desks bore this trimming, because the team had hauled them and the chairs in from the curb outside a neighboring business shortly after the members arrived, to avoid the cost of buying furniture. And somehow it felt right to keep the trash tape in place, at least until they got this scrappy startup off the ground.

As they debated how to get to a more viable version, Stoppelman and the team found themselves cycling back to a previously considered and rejected notion: the idea of placing priority on the *recommenders* as the users Yelp must optimize its interface to serve. Could they turn Yelp into a platform designed to attract review submissions, independent of anyone's specific request? So, what incentive could be dangled to get more people reflecting on their experience with businesses and submitting reviews? One possibility, of course, was to pay for their reviews—this was the trusty model used by long-established publishers Michelin Guides and Rough Guides. Hard to conceive, though, would be the platform's ability to scale up to cover millions of businesses and stay on top of a highly dynamic local landscape. It was also inconceivable that Yelp could fund such an effort. Instead, Yelp would somehow have to do something that had never been done before: persuade a huge volunteer army of normal consumers that writing reviews, unpaid and mainly for the benefit of strangers, was a good use of their time.

By 2004, it wasn't a completely novel concept that sites would serve as platforms for ordinary folks to post their own content. Blogging platforms were already enjoying some use, the earliest of them, Open Diary, having been launched in 1988. These made it possible for people without any programming skill to publish their own "weblogs." Most such platforms not only allowed bloggers to post but also had comment sections for readers to respond with their own thoughts. A site called Everything2.com was already well established, too, representing a kind of proto-Wikipedia, where people contributed entries on whatever topics they were knowledgeable about, from Plutarch to Mario Kart.[3] But nobody had figured out yet how to nudge people to do more content creation—enough to populate a site that needed an impressive level of comprehensiveness to be useful at all.

Max Levchin was a great student of tech businesses and had by that point seen many startups reach an impasse. He had considered how some of them got past those hurdles while so many others folded. This focus on learning from experience was what gave him the confidence that he could help teams execute properly. But what did executing properly mean for Yelp now? Yelp CEO, Jeremy, knew very well from his PayPal days that early pivots were an inevitable part of the journey. He also knew that one cardinal rule was to learn from your lead users, understanding what they valued and doubling down to provide more of it. As the team focused on user behavior, one of Yelp's earliest review writers would soon make a difference.

. . .

"Kevin S." was tall and slim with short blond hair and looked to be in his midthirties. He absolutely loved his city of San Francisco and all its varied treasures. He had a real job—he may have worked as an ad copywriter, he wrote so well—but anyway, he wasn't aspiring to become a professional food critic.

But before Yelp's founding team knew any of these details about Kevin S., Jeremy had spotted this review writer as the kind of user Yelp would love to clone a million times. "Terminal access to code base" is something Yelp's software engineers all had during the first few months. So from the first day that the site began inviting business reviews from all and sundry, Jeremy could see the reviews coming in—content miraculously being user-generated—in real time. Anytime someone posted a review, it appeared at the top of his screen. Since the launch day, October 20, 2004, Jeremy had been directly querying the database using SQL and noticing a surprising phenomenon—not only were users submitting a single review, but they were writing sequential reviews! To picture the scene, imagine a black screen overlaid with bright green text, as seen in *The Matrix* or, for that matter, *WarGames* or any other 1980s movie involving computer programs. As Jeremy was in the database and watching this slow trickle of submissions, he smiled at the amusing, spot-on judgment rendered by one restaurant diner, only to see a review of a different restaurant come in next under the same username and then another soon after. "Hey," Jeremy called out, "come here and check this out." The rest of the company, all four of them, gathered round.

Now they were all watching and reading as Kevin S. entered more reviews, each more peppery than the last.[4] Obviously, the guy was a massive foodie. And his comments were very, very funny. Every few minutes, a new one would hit the database and get everyone hooting with laughter. Someone called out, "These are brilliant!" But along with the delight came bafflement. "What is going on here—why is he writing one after another?"

In fact, it took the group a minute to figure out even *how* Kevin S. was posting his free-form reviews in such rapid succession. Yelp's interface used a question/answer flow to make it easy for people to create a decently composed critique, but after the last question, chief technology officer (CTO) Russ Simmons had speculatively added a link, easily overlooked, inviting users to "write another review." Clicking this link was giving Kevin a text box for capturing further reviews, bypassing the designed pathway.

So that solved the puzzle of how Kevin S. was posting so prolifically, but it still left the question of why. Note that there were no profile pages in those days, so it's not as though anyone could really become known for their Yelp oeuvre, with their whole body of work visible and accessible. Kevin must have known he was not building a reputation but was just posting one-off reviews that would be seen by different searchers. Even if readers admired his style, they had no way to seek out more of his critiques.

Is it shocking to learn that Yelp did not have profile pages? Again, cast your mind back to that primordial past of the social web when user-generated content was not yet a thing. This was at the very start of that wave, which would soon disrupt whole industries by allowing ordinary folk to become content creators and even influencers, assembling unique bodies of output with their personal stamp on them. This was 2004, the year that would see the arrival of protosocial networks such as Bebo, Myspace, and Xanga, with their varied approaches to interface and navigation. The reason everyone had gathered around Jeremy's screen was that a user was doing something strange, not something the designers of the newly launched Yelp site had envisioned.

Jeremy had assumed there would be a need for pivots from the onset, and early adopters like Kevin S. signaled to him the type of user experience that was needed to gain wider engagement with Yelp. The insight was to pivot the site design toward a complete relaunch in 2005 to encourage more of this repeat behavior. To find out more about what drove Kevin to leave multiple reviews, Bob, as product manager, contacted Kevin to ask for a chat, and a phone conversation with the mysterious reviewer ensued. Kevin explained that he had created one submission (in response to the original Q&A system the site provided) and enjoyed it so much he wanted to write another. When Bob noted that, in fact, Kevin had written twenty-two reviews that day, Kevin laughed and said, "I guess I just sort of got on a roll with it!" Earlier that day, one of the Yelp team had marveled, "He seems to be writing for the fun of it!" and here was the confirmation of that observation. But could it really be that simple? For a team that had been puzzling over what form of compensation might induce people to spend time entering reviews, it was an epiphany that people, at least some people, actually just like writing and sharing their opinions. The way to double down on this conclusion would be to make the experience even more fun for them, and there must be ways to do that. This was part of Jeremy's aha moment that put Yelp on the path toward generating the millions of reviews it needed each year.

Soon after this, Bob and the rest of the Yelp team met Kevin S. in person, too—an easy-enough meeting to arrange, given that he lived in town. It was a casual conversation on one level, with a lot of laughs, but also very important in shifting the team's mindset: the developers had been aiming to serve the person seeking recommendations, but now there had to be a fundamental reorientation to focus on the content creators. The questions directed at Kevin were all about what could keep him on that roll. And more broadly, how they could make a site where it was just very attractive for people to write their content. And here, Kevin made his next contribution to the future of user content creation. One thing he said he'd love was to be able to go back and find his own reviews.

Making someone's own content findable—what a concept! There were plenty of notes from that meeting, but this was the need that got the team thinking hard, and it ultimately turned into the idea of creating individual profile pages. If Kevin S. had a dedicated page, it wouldn't only serve his archival needs. It would also be a place where other users could learn about him and see his reviews in one place. Maybe there could be a way for individuals with their own pages to connect with each other (to add "friends") and exchange messages. Jeremy came to the conclusion that moving to more "structured" blog-like profiles for users with chronological reviews would encourage the desired behavior of users writing reviews for the fun of it. Within a few months, fueled by ideas like this, the team had completely redesigned the site to incorporate these profile pages and rebranded the site with different colors, a new interface, and the modern logo designed by Yelp's graphic designer, Michael Ernst. Thus, in February 2005, they launched what was internally called v.6 of Yelp. It's essentially the design still in use today.

At this point, with a site that everyone at Yelp agreed was better, it was a good moment for additional in-person conversations with people who, like Kevin S., had used the site with some regularity and posted multiple reviews. That wasn't going to be too hard: there were only twenty or thirty of them, and Nish Nadaraja as marketing manager had already been in touch with each of them by email at one point or another. Small customer base that they were, someone at Yelp realized it would be more efficient to just get them together to talk about the product. Yelp had never done a launch party for its previous iterations, so that's what the team called the gathering, which they hosted at Emporio Armani. In fact, it was an early version of a users' event, and while it was a little awkward at first since nobody had ever met before, people soon found enough to talk about. Stellah D. was one of the users who attended. She was an artist with a distinctive and edgy style, a proto influencer. As Jeremy chatted with Stellah, he realized that, like Kevin S., Yelp's key demographic at the time was more "urban early adopter" than the typical techie early adopter.

This became more clear the day after the event, when the usage of the site, surprisingly, went way up. No one at Yelp had expected this, or they might have deliberately planned for it. They had just been hoping for some user feedback. But evidently, enough social connections had been made at the party that a lot of their guests were finding each other's profiles on the site. And as they read each other's reviews, they were indeed starting to use the messaging feature. Many of the brief notes they sent included compliments on each other's reviews, and that positive feedback seemed in turn to be spurring people to write new reviews. These were clearly social, culturally savvy users who wanted to engage in Yelp's online community, and Jeremy realized the team should be focused on building a brand and interface that would appeal to them and then the masses would follow.

It was another aha moment.

. . .

Seeing the uptick in submissions after users started responding to reviews and complimenting each other via messages, the Yelp team realized that a key to spurring more content creation would be encouraging more people to do this. They decided to add a "send a compliment" feature. Russ Simmons asked Bob to come up with some ideas for how this option would appear on the screen.

Bob did some research, starting by looking around the neighborhood: What features closest to this idea were already being tried on the consumer web? Most sites at the time were still not social in nature but were dedicated to e-commerce and information publishing. Still, some sites had features the Yelp team could learn from. Most prominently, Amazon, which had been inviting customers to post written reviews of products since 1995, had recently added a way for others to show appreciation of this input by clicking on a blue hyperlinked phrase: "Like this review." It was a crowdsourced way of driving the most helpful product reviews to the top of the heap.

A site called Delicious (del.icio.us) had recently been launched as a kind of social resource to help people discover, collect, and share websites of interest. Since the first days of the web, people had been maintaining bookmark lists of URLs that they usually annotated with reminders of what they liked about a site. Delicious took that habit into a new realm of social bookmarking by allowing people to combine their finds into different categories—in a way preparing the ground for Reddit, the platform that would come later for communities to form around shared hobbies and passions. A Delicious user could click a thumbs-up icon next to a URL contributed by another user. It wasn't a way of rating the original poster's contribution but simply the mechanism for adding that URL to their own set of bookmarks. Still, it was an existing model for how to indicate a like.

There was also HOTorNOT (originally called Am I Hot or Not?), launched in October 2000 by James Hong and Jim Young. The point of the site was for users to upload photos of themselves for the rest to rate on a one-to-ten hotness scale. We'll say more about this particular startup in chapter 2, but the point here is that although this was not yet a full-fledged like button, it was an emotion-laden response to a piece of user-generated content.

And even earlier, another kind of digital product had taken off and was teaching many thousands of early adopter types to vote content up and down. This was TiVo, which in retrospect we might describe as a midway point between the VCRs of the 1980s and 1990s—those stand-alone recorders and players of VHS tapes—and Netflix and other streaming services that would take over after 2007. TiVo was a digital video recorder that allowed its owners to forgo physical cassettes; it also offered up a large inventory of preaired television programs users could select.

We asked Jeff Dodds, the marketer responsible for Virgin's launch of TiVo in the United Kingdom, what he recalled about that service and the handheld device that was used to control it. To start with, he reminded us that before TiVo's US launch in 1999, no television watchers had the ability to access TV shows other than those they had themselves chosen to record on their VCRs. He noted that TiVo was especially revolutionary in its use of software to serve up recommendations based on a user's past use, "rather than you scrolling endlessly, looking for things you might like." The other point of great interest to us was that the TiVo remote control device needed to come up with buttons for users to press, depending on whether they wanted to add programs to their queue or skip them, and those buttons were in the shapes of little hands—on the right, a green thumbs-up, and on the left, a red thumbs-down (figure 1-1). Given the product's huge success across the next few years, its interface was certainly part of the context that influenced Bob and other user interface (UI) developers around 2004.

Finally, before any of these developments, voting had even been possible on the bulletin board systems—known as BBSs—that, beginning in the 1980s, gave personal computer owners their first access to networks and gave rise to an active subculture of content posting, until they were dramatically eclipsed by the arrival of the public internet. It was in hindsight a ridiculously limited system, in that the typical user could only dial in and post one message, after which they would have to hang up so that someone else could dial in to post another message. But somehow, even under such constraints, through an MS-DOS operating environment called Renegade, BBS offered rudimentary discussion groups, messaging functionality, and a voting feature.

FIGURE 1-1

TiVo's "peanut" remote control, designed by Rick Lewis of IDEO

Source: Courtesy of Evan Amos

For Yelp's UI design challenge, the point here is that there were other instances of digital platform users being given a way to register approval of something, but there was nothing just like what Yelp was trying to do. The important thing for Yelp's purposes was to intensify that incentive loop the team had gotten a glimpse of. It wasn't enough to give users a way to like a piece of user-generated content. The user who generated that content also needed to feel that affirmation personally. It needed to be a compliment both given and received. And that meant Yelp would have to invent its own solution.

In the version Yelp first created, if you posted a review, anyone reading it had a visible way to pay you a compliment if they thought it was good: they clicked a link for one of the different compliment types offered—choosing, perhaps, *You're cool* or *Hot review.* Clicking the link would open a text window with a boilerplate compliment they could replace or edit as they wanted. (Bob's signature from working on this at Yelp is that, to this day, the example name used in the template, and Find Friends search box, is "Bob.") Then, after that message was posted, you as the review writer got an email. You clicked through to read the compliment, and then the compliment appeared on your profile. The hope was that you would feel so good about being complimented that you'd write more reviews.

It was a very popular feature, and Jeremy and Russ quickly noticed a strong correlation between the number of compliments people received and the number of reviews they subsequently wrote. Soon they were coming up with more compliment types and linked phrases that people could click on, doubling down on a feature that worked like a charm.

But it still wasn't good enough, because there was still too much friction in the process. To send a compliment, the user had to click a link, edit the content on that page, and click again to submit it. Three steps meant two page refreshes—a real drag in an era when Yelp's consumer customer base was still largely dependent on dial-up internet access and when a refresh typically meant three to five seconds of delay. After a few months of watching positive feedback drive content creation, Russ Simmons put the question to Bob: "How could we make this even easier to use? How simple could we make it to send a compliment?"

. . .

Bob is an inveterate diary keeper. In fact, he's a hoarder of information related to his life. And while it's not so unusual today for someone to have, say, a ring that records their heart rate and so forth—as Bob does—this habit of his goes back to his boyhood in the predigital age. "I've always been obsessed with collecting information," he explains. "I saved every receipt, every bus ticket. I had boxes and boxes." One time around 2010, a friend came to his house and saw the manifestation of this penchant, and Bob told her how it had gotten started, back in 1992. She said, "This needs to stop. I don't think this is healthy." But the thought of documenting his whole life had taken hold when he was a child, even if he had no real sense that he would need the information—and in fact had never used any of it. He just had some sense that it might add up to something or be useful to mine someday.

If it weren't for his documenting habit, we wouldn't have an artifact of the earliest days of the social media like button and no one would remember the day that artifact was created—because, really, it did not feel like a moment that had to be preserved for posterity.

The question Russ raised about making Yelp compliments easier had two parts. One was a technical question, and the other was a UI question. Russ, who was not only Yelp's cofounder but its chief technology officer, had the technical challenge to take some clicks out and make the process as streamlined as possible. How simple and fast could it be? He thought they might even get it down to one click with some creative use of JavaScript, the programming language Netscape had developed to add dynamic elements to web pages. JavaScript had never been used for a purpose like incorporating user feedback. The hope at Yelp was to allow a user to register a response while staying on the same page, avoiding the change in URL that would force a page refresh—and keeping them on the page where they could see and might respond to additional reviews. Being aware of new applications that used JavaScript for a more responsive user experience, such as Gmail, Russ figured that this capability should be possible but that maybe there was a reason it wasn't. As soon as the team members discussed this application, its potential seemed so obviously valuable that they were surprised it didn't exist on the web already.

Meanwhile, Bob had the interface design part of the task. His starting point was the same question: Could it just require a click on a page of content—a click that didn't take the compliment payer to another page but simply updated a *counter* on that page? And if so, what would it need to look like to be as simple as possible for the user? And this is how he found himself one day with a notebook in hand, thinking about different options for the design of that clickable thing and sketching a hand making a thumbs-up gesture, with the number of likes next to the button. It's the drawing you can see today in Martin's online museum of imagination, the Napkin Gallery.[5]

UI design at the time wasn't the well-studied process it is now. But the work of optimizing the user experience was already evolving into a discipline. Bob, Russ, and the handful of others at Yelp were all well versed in the book then regarded as the bible of web design: Steve Krug's *Don't Make Me Think*.[6] Krug, a veteran of many website launches and revamps, published this manual in 2000 to share the most important principles of designing for consumers. These principles could be summed up as *radical simplicity* and *intuitive ease of use*. The title of the book was what he declared to be "the first rule of usability." If you design a feature that forces the user to stop and think before using it—perhaps because it isn't clear where to click or what will happen after a click—then you undermine the performance of that feature. The interface has to be as intuitive as possible; the flow needs to be as friction-free as it can be.

When Russ Simmons declared that Yelp had to streamline its compliment process and take some clicks out, he could have been quoting directly from Krug's book. And when Bob sketched out some options, he was acting on other advice Krug emphasized: a good way not to make people think is to hijack impulses they already have and to employ imagery that already resonates with them. This meant scanning for what was already out there on the web making some ways of expressing emotion online more familiar than others. But it also meant reflecting on how people signal enjoyment of things they encounter in their offline lives. For someone drawing on a recent background as a medievalist, it even meant delving into how people had reacted to text in the past—for example, the fifteenth-century monks' addition of manicules to the manuscripts they spent their days studying, to draw attention to passages they particularly liked. It was in the midst of this hunt for time-honored symbols that Bob was prompted to write the word *like* and sketch a thumbs-up with a counter.

The truth is that the day Bob made this sketch did not feel in any way significant. It was just that day's task to come up with a streamlined compliment button. And another truth is that when Bob shared this sketch with his colleagues, he presented it as just one of various options. What's more, the radically simple thumbs-up icon wasn't actually the direction the group decided to take. Everyone agreed that people responding positively to reviews had different reasons for liking them and that the why of the like was useful information to capture.[7]

So in the version of Yelp that went live in May 2005 (v.10), it offered a choice to click any of three buttons: *useful*, *funny*, and *cool* (figure 1-2). When you went to a review page, you noticed a panel including these choices for a reaction, and if you wanted to compliment the writer, you simply clicked one reaction. That click instantly added to the counter and triggered an alert to let the writer know they were getting more props. Simmons had created the JavaScript widget, while Bob and others had designed the interface for what would be the first one-click mechanism to choose from

FIGURE 1-2

Screenshot of Yelp's original *useful*, *funny*, and *cool* options

Was this review …?　Useful ● (7)　　Funny ● (5)　　Cool ● (6)

Source: Courtesy of Yelp

one of multiple emotional reactions to another user's content. And thus, in May 2005, emotional reaction buttons as we know them were born.

. . .

The effect was immediate and impressive. In a recent conversation, Levchin recalled that before that moment, everyone knew that "just 1 percent of people would write and create content that people actually would read." Yelp's big achievement was that it exploded that percentage. "It was the first time I had seen an attempt to recruit some of the 99 percent with some super-low barrier to entry for content creation. That was the idea." In short, he said, "the psychological or psycho-behavioral foundation of the like button is really all about breaking out of the 'only 1 percent who will say anything online' assumption."

It would make a tidy story to claim that from this tiny seedling grew the mighty tree that is today's like button usage, but in reality, many other seeds were being sown around Yelp's social web neighborhood, too. Yelp wasn't the only startup with a business model predicated on user-generated content and not the only place where it was dawning on developers that peer-level reactions to that content had a big role to play.

Take Digg.com, a news aggregation site launched in 2004. The site would later go through various overhauls, but from the beginning, the idea was to let people share things they found on the web. Digg users submitted pointers to news articles and other information accessible online, giving each submission a headline and synopsis to help others determine if the content was worth their time. The name was clever because, first, someone was digging up something to promote and then others were deciding whether to "digg" it (vote it up) or "bury" it (vote it down). If a story got a certain number of diggs within a specified period, it appeared in the "popular" section of its category—and really popular stories were featured on the home page. If enough people clicked "bury," the content would disappear from the site's main feed (and be served up only if a user specifically chose to "include buried stories" in a search).

Or consider Everything2.com, which had added its own way for users to rate each other's content. As mentioned, this was a site where people could post encyclopedia-style entries on any topics they felt deserved attention. Its designers went on to pioneer what became known as the experience point (XP) system: users won XPs for engaging in different actions useful to the site's success, chief among them contributing useful entries. For the user, the value of accumulating these was that higher XP status conferred certain privileges. Only users at a certain XP level were allowed, for example, to pass judgment on other contributors' content, which they did by registering up or downvotes on writing quality and dispensing "cools" with a C! button. The voting had real teeth to it, since too many downvotes would get a post removed from the site.

Similarly, the early blog platform Xanga introduced what it called eProps in December 2000. This small blue link appeared beneath each user-generated article and contained the number of votes given to that blog. And the platform sent an email to the content creator with a link to the profile of the user who had shared the eProps. The person behind this feature at the time was Xanga's creative director, Biz Stone (who went on to help popularize blogging at Blogger and podcasting at Odeo and then cofounded Twitter). When we asked him about eProps recently, he said, "The original thinking behind it that I had was, if you're too lazy to leave a comment, you could just hit this button. And that means 'I acknowledge that I read your thing.'" Stone was so intrigued by the power of this feature in action that when Google acquired Blogger in 2003 and he found himself with a bit of time on his hands, he wrote an internal memo and distributed it at Google proposing a social system that he called SMURF: Single Multi-Use Rating and Feedback System.

And then there was Vimeo, a video-sharing site that in 2005 added a truly straightforward like button. In fact, *Fortune* would later highlight its November 17, 2005, introduction of the feature as "a day that lives in online infamy . . . when the first 'Like' button was clicked."[8] Developer Andrew Pile freely admitted to *Fortune*, however, that the button had been a slightly modified imitation of Digg's up-vote feature: "We liked the Digg concept, but we didn't want to call it 'Diggs,' so we came up with 'Likes.'"

If we wanted to parse the differences in the progenitors and early versions of like buttons, we could examine each of them with a simple question set: Did it enable expressions of a single emotion or multiple emotions? Did it feature instant updating without reloading the page? Did clicking the button add anything to an individual profile of the person whose content was liked? In other words, was there an underlying social graph? Eventually, all these aspects came together in what is now thought of as a like button, but features that didn't do all these things were important steps along the way.

It only makes sense, in fact, that there would be variations on the button because different companies had different business reasons driving their decisions to incorporate such a feature. A mechanism allowing a user to say they like (or dislike) something can perform a variety of functions: it can add up to crowdsourced voting, give users an emotional outlet, create a basis for interpersonal engagement, provide valuable feedback to companies, and more. Amazon, for example, arrived at the idea because it wanted to know which reviews should be amplified—moved up in the stack—to increase the likelihood of a satisfying purchase. For a business aiming to be "the most customer-centric company in the world," the like button was also a visible way of putting customers in the driver's seat. (Interestingly, in December 2020, Amazon retired the feature that had allowed comments on other customers' reviews, saying that "the comments feature on customer reviews was rarely used."[9] It might also have been that the function had devolved into sellers defending themselves against customers' critical comments. Today, Amazon provides only the options to deem a review helpful or to report it as offensive.)

For Yelp, the reason for creating the like function was not to facilitate online buying but to encourage more review writing. This is why the team didn't offer any "dislike" options. Digg had yet another goal, to elevate the most popular things to read and ditch the awful stuff. And when Facebook ultimately added the like button in 2009, a big part of the business case was ad sales. In the words of Justin Rosenstein, product manager of Pages, "You could create advertising that was tied to someone liking a particular page, or you could advertise a page to people who liked similar pages."[10] Depending on what it was trying to achieve with the like, every site was putting its own spin on its programming and interface.

· · ·

Even as all this work happened in pockets, no one was really doing it in isolation. Because the culture of innovation in Silicon Valley involves so much interaction, collaboration, and idea sharing, people were watching each other's moves to add emotive reactions to their different site designs. And thus the like button was subject to rampant borrowing and adaptation and, over time, patterns of divergence and convergence.

A fabled aspect of the Silicon Valley culture is that on top of the pollination effects of workers constantly jumping from place to place, there are frequent meetups where people freely show off their work, compare techniques, and trade ideas. For people working in the realm of UI and UX in the early 2000s, the greatest nexus for this activity was the annual meetup organized by Tim O'Reilly and Sara Winge and known as Foo Camp (a play on a commonly used placeholder programming variable *foobar* and the backronym Friends of O'Reilly). When it launched in 2003, Foo Camp was still a small community of designers and developers, with around two hundred attendees in its early years, as Bob recalls. Hosted at the offices and the O'Reilly home, it was one of the first "unconferences" in Silicon Valley. It took pride in being self-organized, prioritized the open sharing of ideas and problems, and made connections in an informal setting (a campsite, where people were often sleeping in sleeping bags on the office floor). Arwen O'Reilly, Tim's daughter, along with her husband Saul Griffith, also cohosted a monthly gathering at Squid Labs, in the East Bay region of the Bay Area.

After Yelp launched its compliment feature offering its choices of *useful*, *funny*, and *cool*, the team noticed the same kind of feature being added to other sites. Other ideas were radiating, too, for how social site users could display reactions. Foursquare, a pioneering platform in location-based services, made the concept of badges central to its user experience. It capitalized on GPS technology to allow smartphone users to check in when they arrived at some defined location like a restaurant or museum, often to alert their friends to their movements. The site also served up recommendations of other popular spots in the same vicinity; users could add these places to their to-do lists. Visiting a spot known to the system earned the user a badge—and collecting badges on profiles proved so appealing that the feature started to crop up on other sites.[11]

Another example was when the concept of gamification (turning a set of functional actions into a game by using points, levels, competition, playful language, and so on) started to be embraced.[12] Yelp was using that term very early, and soon enough, the Foursquare folks were talking about it, too, generating more interest and excitement and taking the ideas to the next level. They helped popularize the concepts of gamification and encouraging users to repeat certain behaviors on the social web. The list could go on and on—there are endless examples of how this contagion played out. Once a Foursquare, a Yelp, or another shop that was recognized and admired by the development community in San Francisco started experimenting with a new concept, other designers who wanted to be on the cutting edge started to play around with the ideas, and everything proliferated and advanced quickly.

. . .

The common belief that Facebook invented the like button is both mistaken and quite understandable. This social network was hardly the first to have a one-click positive emotion response or even to label it "like." But Facebook did create an iconic form of this response and, by putting it at the fingertips of its massive user base in 2009, caused liking to scale up enormously.

It's a little strange, actually, that Facebook can't claim to have invented the like button. In its germinal form as Facemash, the website that Mark Zuckerberg unleashed on the Harvard student body was purely an invitation to users to register a reaction of a somewhat emotional nature—choosing which of two randomly served-up classmates was the more physically attractive. And when Facebook launched to the public in 2004, it was a platform for user-generated content via profile pages that provided for interaction, including the ability to "poke" another user by clicking on an icon of a hand. Yet Facebook did not decide to add a like button till fully five years later. And this decision was not because the company was too distracted by other matters to come up with the idea. It was a deliberate and repeated decision by Zuckerberg not to add a feature that was internally proposed but that he deemed to add no value.

It appears that it was another startup, founded by Paul Buchheit and Bret Taylor and later acquired by Zuckerberg, that brought the feature to the Facebook table. "FriendFeed was the original like button," Buchheit said, pointing to its inclusion in a site design that went live on October 30, 2007. This was two years before the acquisition, but Buchheit (who is also credited as the creator of Gmail) recalls that Facebook had been keeping an eye on his company as a place to learn from and potentially to buy.[13]

We checked in with FriendFeed's founding team to hear their various memories of what they had in mind when they created their button, which featured the word *like*, an icon of a smiley face, and the names of users who had clicked it. FriendFeed started out as a social media aggregator—a place where a person who maintained separate accounts on various platforms (say, Twitter, Flickr, and YouTube) could see the activity across all of those, compiled into one unified feed. It also allowed for commenting on FriendFeed itself. One challenge in the early days was what Buchheit calls the "empty room" problem. "One of the worst things on a social network is if it feels dead," he said. "The feeling of talking into an empty room is demotivating. Every time you post something, you think you might as well be throwing it down a well." The key, then, is to get more users responding to posts—but that is also a challenge. When someone shares something, Buchheit said, a lot of people will think, "It's great that it was posted—but, well, I don't really have anything to *say* about it."

The solution was to create what the team called "one-click commenting"—a tap on a like button that would convey meaning and, importantly, that would come from a known account. For the responder, that setup would drop the threshold dramatically, Buchheit explained: "You don't have to think of anything to say—you just click the like button." And for the original poster, the fact that a name is attached to that comment—that you see that *Paul likes your post*— would make it much more socially meaningful. This was, Buchheit said, "the thing that really, to my mind, distinguished the like button" from the upvoting and downvoting that was already common on sites like Reddit, YouTube, and Digg: "It's not a vote. It's a one-click *comment* and it's *attached to a person*."

Better still, that attachment to a person meant that FriendFeed could connect a like to the knowledge it was gaining of your social connections, as it created the "social graph" of your network and used it to customize your display. "We would call out the likes of the person who was closest to you," Buchheit explained. "If three people liked a post and one of them was your friend, we would list your friend's name first." The benefit was to serve the user better, because "if someone you know likes a thing, that's more meaningful than if some rando likes it."

And therefore, the button provided a natural way of expanding distribution, through what the FriendFeed team liked to call *friend-to-friend content*. Buchheit explained: "Say there's someone I'm not following, I'm not subscribed to, but my friend likes something they posted. That thing can then be pulled into my feed based on that." He was describing an action that by now is very familiar to social media users, who all expect to see content served up that "you might like." But in 2007, this would not have felt so innocuous—it might cause a user to wonder, "Why am I seeing this random post?" The like button got people past that confusion: "OK, so, it's because my friend liked it." And therefore a like, in Buchheit's words, "was providing that signal to the engine that it can recommend additional content." No further explanation would be required to explain why a user was getting something they didn't subscribe to, or from someone they didn't follow.

Buchheit was quick to give credit to his colleague Bret Taylor for actually building the feature. People like to talk about great software engineers as 10X performers of their art because they are capable of producing ten times the amount of good code than their peers produce in a given day. Taylor, as Buchheit described him, is a "1000X engineer." He created the new button in short order. But Taylor in turn tipped his hat to Ana Muller for the fact that the button he created was labeled *like*. He remembered that the first impulse on the part of the product developers was to use a heart icon, in which case, of course, it would be little expressions of *love* being dispensed, not just likes.

Muller remembered the day well. It was a small team—besides Buchheit and Taylor, there were Jim Norris, Sanjeev Singh, and a couple of other software designers—and "we all sat within five or six feet of each other," she said. While her role was to serve as CMO (as she says, "Chief Miscellaneous Officer") and not work on product, it was impossible not to overhear their conversations. And when she heard her colleagues talking about the button and convincing themselves that a heart was probably the way to go, her reaction was so visceral she had to speak up. "I believe," she told us, "my exact phrase was, 'I refuse to work on a website that has hearts all over it!'" It was, she said, "an unequivocal *no* . . . a line in the sand" on a move she could barely believe the group was even contemplating. "I distinctly remember saying, 'What is this? You're a group of hardcore engineers. How am *I* the person saying we don't want to have hearts all over the screen?'" And in an interesting parallel, when we spoke with YouTube cofounder Steve Chen, he let us know that YouTube did in fact start out with the heart icon, but "within a few weeks, we switched to a thumbs-up icon." To this day, Muller says she can't really articulate why she spoke up so adamantly. Maybe it felt too childish, or frivolous, or unprofessional—but it wasn't as if Friend-Feed was a stuffy place. She herself was the one who came up with the idea to launch the "AirGrievance" tool (a new way of interacting with FriendFeed via the creation of native posts)—aptly enough on Festivus, the mock holiday from *Seinfeld* known for its "airing of the grievances." But her feeling was that a one-click love would be cringe: "I think, honestly, it was so obvious to me that this was a bad idea that I couldn't even explain it."

The choice of the button didn't seem like that big a deal to the others, so the matter was settled. Buchheit thought he might also have been swayed by the name of a startup he had recently invested in. It was a casual-dating site called I'm in Like with You (later, iminlikewithyou.com would pivot away from dating to casual gaming and rebrand itself as OMGPop.com, whereupon it was acquired by Zynga). In any case, the decision was made. It would be a like button. And its icon? "We used a smiley face," Buchheit recalled, "and I just pulled that out of one of these icon packages. It was free.

"Obviously when we launched it," he said, "we didn't realize that it was going to be a thing that people would write books about. It seemed like a relatively harmless feature." But in retrospect, Buchheit thought labeling the button "like" was a big part of its success. With the emphasis on keeping the threshold low for someone to register a comment, its mildness and versatility were well suited to the problems his team was trying to solve. Buchheit summed it up: "The history of the like is pretty simple. We had a little social network with not very many users. And the simplest version of it, as we described it early on, was one-click commenting."

· · ·

But Facebook engineer Andrew Bosworth takes issue with the implication that Facebook outright copied the idea. In a response to a query posted on the website Quora, he presents a timeline showing an internal communication about a potential like feature on August 22, 2007. Another page from Bosworth's notes documents that engineers were ready to launch such a button by November 12, 2007—making it likely that the notion had occurred to them before they learned of FriendFeed's like button in development. Steven Levy tells more of the story from the company's perspective in *Facebook: The Inside Story*. After an internal designer named Leah Pearlman suggested in July 2007 that Facebook should offer a friction-free way to indicate enthusiasm, Levy writes, a working group got to work on what it decided to name the "Awesome button," which would use a star as its clickable icon.[14]

The one thing that is abundantly clear is that Zuckerberg didn't want the launch to happen. As Bosworth wrote in his work diary on that day in November: "Final review with Zuck surprisingly doesn't go well. Concerns about whether the interaction is public or private, cannibalizing from the share feature, and potential conflict with Beacon. Feature development as originally envisioned basically stops." The button was paused and wouldn't get the thumbs-up until seven internal reviews, one unauthorized experiment, and fifteen months later. Facebook finally launched it on February 9,

2009—and styled it as a like button over the objections of the development team. Recalled Facebook engineer Tom Whitnah, "We were all stubbornly insistent that no word could be more awesome than 'Awesome' and Zuck was the main person to recognize it wasn't a good choice."[15]

Bob recalled seeing the feature that day and thinking, "Wow—that is really clean and well done." The design, by Facebook's UI designer Aaron Sittig, featured a hand giving a thumbs-up signal.

After Facebook finally added its like button, the feature proceeded to spread like wildfire, both in its use and through its replication on other sites. It was a watershed moment. The like button was so significantly popularized by Facebook's adoption that we might think of two like-button eras. There's the *pre*-Facebook history of the like button—the history to which we've devoted most of this chapter—and there's the *post*-Facebook history that most social media users today have an awareness of. It was like an avalanche of copycat adoption picking up pace and spreading beyond the social internet to commerce sites, media sites, and videoconferencing platforms.

Facebook even made propagating the like button part of a new revenue stream for itself. Launching what it called its Open Graph initiative in 2010, Facebook became a vendor of plug-in technology to web startups, allowing them to avoid reinventing the wheel by simply adopting its own well-developed like button and other features.[16] It was a big move and put Facebook into competition with other plug-in purveyors, but the initiative showed the company's strategic creativity: adopting Facebook's like button as a plug-in essentially turned other sites' pages into Facebook fan pages. When someone clicked on the like button on your site's page, they became a fan of the page and the link was shared with their network on Facebook, too.

The like button soon became almost obligatory—conspicuous by its absence on the sites that did not include it. YouTube, for example, added a like button in the form of a thumbs-up in 2010 even though the feature did not yield data the site felt it needed to decide what to promote. Since the beginning of its video-sharing service, You-Tube has tallied actual views of content and allowed those to drive individual videos up or down in visibility. The addition of likes (and dislikes) was a response to users' acquired belief that they should be able to render judgment on what they had just taken the time to watch and, in the process, display their own tastes.

Other holdouts that finally capitulated include Twitter (since re-branded as X), which could fairly assume that a retweet was enough of an affirmation. And for good measure, the site also allowed its users to "favorite" tweets posted by others. In November 2015, it changed the look of the latter function, retiring the former gold star icon and replacing it with a red heart, while also renaming it as a like rather than a favorite. Its announcement of the change explained that it would alleviate user confusion and put the function more in line with other social networks.[17] When we spoke with Twitter cofounder Biz Stone recently, he explained that the concept of fol-lowing, which he said was coined by Twitter, "was a sort of pseudo-liking in itself." Bob looked up his own invite email from Twitter (then called Twttr), received from cofounder Evan "Ev" Williams on July 14, 2006, and the idea of following was central from the very start. The email reads, "Once you get all set up, ev and you will auto-matically get updates from each other. This is how it works for any-body you add as a friend from here till the end of time . . . well . . . not really." In light of this clunky invite email, it was not clear to anyone that, as Bob signed up as user number 920, hundreds of millions of users would follow.

LinkedIn was one of the last major online networks to add the like button—as late as 2012, a full three years after Facebook. When we recently caught up with LinkedIn founder Reid Hoffman, he explained, "In the early years of LinkedIn, we refrained from add-ing features that we considered social, such as profile photos and allowing users to post content. We wanted to establish LinkedIn as the professional space on the internet and were concerned that such features could distract from that. We were building the network by capturing business connections for jobs, sales, and expertise. As LinkedIn evolved, we learned that being a place for business influencers and thought leaders to share their ideas, and allowing members to follow, comment, and like such content, can drive good engagement while staying within the business and professional context." In April 2019, LinkedIn expanded its like button to allow multiple emotions including like, celebrate, love, insightful, and curious.

Instagram was another late adopter, in November 2016, allowing users to like comments on pictures and to tap "like" on a live video broadcast. But immediately, the ability to react to content became important to the site's appeal. (Perhaps *too* important: in late 2019, Instagram dispensed with the like counters on posts in response to concerns that such visible popularity scorekeeping was a source of too much anxiety for some users.) In 2022, Instagram took the bigger step of enabling likes on stories themselves. WhatsApp, too, launched its like button in 2022, after Facebook's acquisition of the company. And when iMessage, from Apple, started to allow users only a couple of years ago to like a message, the general reaction was, What took you so long?

.   .   .

From 2016 forward, Facebook was also responsible for another form of spread in the like button—an expansion of its range of emotional options. In that year, the site parsed liking into five more-specific reactions: *love, haha, wow, sad,* and *angry* (but notably, still no *awesome*). Each had its own icon—which for extra sizzle was now not a static image but an animated one.

LinkedIn followed suit in 2019, expanding on the like button to allow for six possibilities: *like, celebrate, love, insightful, curious,* and *support.* And so a new wage of contagion took hold, with innumerable other sites adapting to serve user bases that had become so completely comfortable with liking that they were ready to branch out to expressing more-nuanced reactions.

Yelp's users, meanwhile, had been using their three options all along (since 2005)—*useful, funny,* and *cool*—and the site's designers could take a certain satisfaction that the world had finally come around to the decision they had reached years earlier, not to try to cram all of their users' positivity into one thumb-shaped box. (Only in April 2023 did Yelp change its reaction options. They are now *helpful, thanks, love,* and *oh no.*) They could also note some additional convergence on another of their early decisions: not to encourage negativity. YouTube, for example, had always had both thumbs-up and thumbs-down buttons, but in 2022, it removed the displayed count of the dislikes. At the same time, it animated the like button (but not the dislike one), presumably to pull users' eyes toward the thumbs-ups and invite more clicks there. When we caught up with YouTube cofounder Steve Chen, he explained, "We needed the thumbs-down option to give the viewer a way to say to the algorithm, 'Hey, don't show me more content like this.'" He had already left YouTube when the downvote count was removed. "I kind of hit myself in the head when I saw that. Like, yeah, why did we keep that around for so long?"

When we spoke recently, Biz Stone recalled the conversations at the blog platform Xanga when the team was adding the eProps feature in 2000. "My cofounders said, what about negative eProps if you don't like it? I was like, 'That's too mean. I don't think we should do that. That's not going to make anybody write.'" Years after Stone had left Xanga, in 2005, the company had a brief period when it allowed users to give their own name to eProps. This is the only example of such open-ended input that we are aware of. If you browse the Internet Archive for Xanga on February 10, 2005, you can see some colorful user-generated alternatives to the word *eProps*, ranging from the optimistic "8 Holiday cheers" to the gothic "12 Blood covered lullabys." This feature only lasted a few months.

As we write this chapter, X (formerly Twitter) is reportedly testing a kind of dislike button—but is proceeding with caution. First, it would allow downvoting only of *replies* to tweets, not the tweets

themselves. And second, it would work in the same way as the platform's innovative Community Notes, which are crowdsourced fact-
checks of tweets: replies would only be down-ranked if they were
disliked by people who typically disagree on things. As an X engineer explains in *TechCrunch*, simply adding up all the negative votes
would encourage a "hivemind like Reddit"—a platform whose downvoting feature is notorious for its punishing pile-ons.[18]

Between this proliferation of emotions, the spread of the like button to untold sites, and the ever-growing user base of social media,
it's impossible to know exactly how many emotional responses are
being registered at this point, on a daily basis, to online content. But
we can venture a reasonable guess. Figure 1-3 shows our estimate of
the growth over time of likes registered worldwide. As shown, the
clicks now add up to over 160 billion per day—roughly equivalent
to every person on the planet, from toddler to great-grandparent,
clicking a like button 20 times daily on average.

. . .

A recent documentary tells the story of the conception, design, and
deployment of an amazing invention: the Mars rovers built by NASA
to explore Earth's closest planetary neighbor.[19] Drawing on thousands of hours of archival video of planning meetings, test runs,
robot deployment, and, eventually, robot demise, the filmmak-
ers faithfully trace every major step in *Spirit's* and *Opportunity's*
planned journeys and what was learned from them. To watch the
film is to be filled with fresh appreciation of the ingenuity and pas-
sionate persistence that goes into the innovation process.

But not many inventions could have such a documentary made.
As we'll see in chapter 2, most innovations that wind up going far
and having wide impact start out with much smaller ambitions. A
narrow problem needs to be solved in the context of some project, it
just feels like a typical day of work, and nobody gets a camera rolling
to capture the achievement for posterity. So it's lucky that we have

FIGURE 1-3

Chart of estimated annual digital likes rising over twenty years

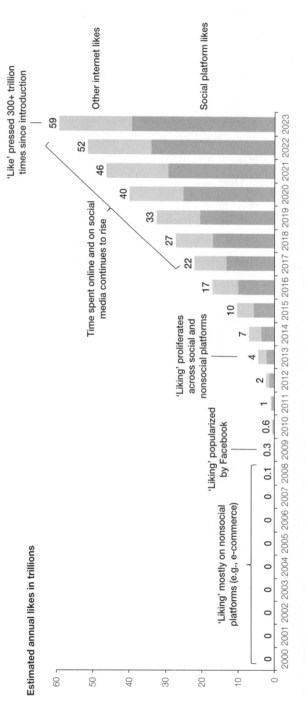

Estimated annual likes in trillions

Source: BCG Henderson Institute research, with sources from DataReportal, Statista, Forbes, Bloomberg, E-marketer, and Everything2

Notes: Annual estimated likes based on number of users, time spent, and rate of likes per minute. "Other internet likes:" includes likes from all social- and user-generated content (i.e., social media networks, social rating and review networks, etc.). "Social platform likes:" includes likes of user-generated content in other platforms (e.g., bookmarking of websites and products, likes on comments in e-commerce websites, but not ratings of the products) and ratings on non-user-generated content specifically made to emulate social media likes (e.g., Hulu's thumbs-up and thumbs-down icons used to rate its video programs).

the twenty-year-old artifact of what may be the first sketch of a like button—and great for Martin that he has an original napkin sketch for his gallery. But the fact that today's most familiar software feature so closely resembles this sketch doesn't mean that the path in between was straightforward. It's a convoluted and contested origin story with various parties able to make a case that they were really the first, and some even bold enough to litigate the matter.[20] It's a case study in messiness, serendipity, and, sometimes, the central actors' cluelessness that they then were onto something so big.

Most of all, the story has yet to reach its conclusion. The like function will continue to evolve in its forms. It will adapt to the needs of site designers focused on new missions and business models. It will exploit next-generation technologies. Two decades into its useful life, the like button is still being invented.

Chapter 2

How Does Innovation Happen?

n chapter 1 we traced the messy process of the invention, evolution, and spread of the like button. This chapter examines how innovation happens more generally, asking if the pattern we saw for the like button is the exception or the rule and what the like button has to teach us in this respect. We begin by delving into two examples of innovation that differ greatly in their degree of orderliness and intentionality. If you're an American of a certain age—let's say, in your forties—and you were anything resembling an early adopter of social media, you will probably remember checking out HOTorNOT, the notoriously sophomoric brainchild of James Hong and Jim Young. Launched in 2000, the website allowed users to upload photos of themselves for the sole purpose of having others rate their attractiveness on a scale of one to ten. Those ratings would average to an overall score, allowing you to know where you stood on an overall peer-determined scale of hotness.

You may also recall that in 2000, HOTorNOT was, well, very hot. Only one week after its coming out, it was seeing daily traffic of close to two million page views and rivaling major news networks as one of the most popular domains on the internet. And in fact, it has managed to age pretty well, too, for a website, still operating today as a dating app under the control of Bumble. What you may not know is that HOTorNOT was also very smart. It wasn't the first photo rating site to hit the internet, but it was undeniably the best in its class—and according to many people present at the dawn of Web 2.0, it created a lot of the basics that other sites would build on. Evan Williams, for example, who would go on to cofound Twitter, described Hong in a recent interview as one of the most intelligent people he knew in Silicon Valley, and Williams gave HOTorNOT props for many of the features and ideas that took hold in the earliest days of the social web.[1] Among these was Young's very clever solution enabling users to react to content without forcing a page refresh with every click of a number. Repurposing JavaScript for a task its inventor never envisioned—much as the Yelp team did for its own purposes—HOTorNOT took all the friction out of its rating process.

So imagine our surprise when we asked Hong to think back to that time—to the moment of devising that early progenitor of the like button—and find him unwilling to take much credit. "Yeah, so, look," he demurred. "I mean, I don't know. OK, we definitely invented it *independently*, but I could never say with certainty that we were the *first*. To the best of my knowledge we were."

He quickly drew a comparison to another invention from the same era: "You know, it's kind of like when Max invented the Gausebeck-Levchin test, but at the same time as Luis von Ahn, who also says *he* invented it." He was talking about Max Levchin's early days as chief technologist at PayPal, when he and David Gausebeck came up with a novel way to fend off fraudsters, unaware of academic work along the very same lines by von Ahn's team at Carnegie Mellon University to produce something they called a Completely Automated Public Turing Test for Telling Computers and Humans Apart—you'll know it as captcha.[2] PayPal's version of the solution was built into its code in 2001, around the time the company was crossing the one million customers mark, and Levchin's deployment of captcha was the first at-scale application of the approach.

As for HOTorNOT's pioneering software, Hong suggested it was just another development whose time had come: "If we hadn't invented it, you know, someone else would have—I mean, like, a lot of smart people out there would do the same thing, right?"

This comment doesn't comport with the way most tales of innovation are told around the campfire, with their heroes experiencing brilliant epiphanies, going for the glory of making their visions real, and knowing that their names will go down in history. So, what are we to make of the contrast? Is it that the like button's convoluted origin story is a special case—an idiosyncratic departure from the normal course of innovation? Or is it a case with broader relevance? What might the like button teach us about how innovation happens—whether specifically in user interface (UI) design or under certain conditions like those prevailing in Silicon Valley in the early 2000s or, in the broadest sense, in all places and every era of human discovery? Once again, our curiosity beckoned us onto an inviting trail, up a slope carved with switchbacks, on our way to getting a clearer view.

. . .

As a contrast to James Hong's rather equivocal account of HOTorNOT's triumph, let's start with a neat story of invention more along the lines we're used to hearing.

Like all such histories, it begins with a clear and pressing problem. The year is 1815, the place is Northern England, and the situation is that coal mining, as essential as it is to fuel the British Empire's rising industrial prosperity, is a very dangerous business. Miners, of course, toil deep below the surface of the land, and their efforts have a nasty habit of penetrating natural pockets of gases, filling the tunnels with what are known as damps. (The term is from *Dampf,* German for "vapor.") A black damp or choke damp is air made unbreathable by excess carbon dioxide, and a white damp has too much toxic carbon monoxide. And then there is the dreaded fire damp, in which methane builds up to the point where a mere spark can set off an explosion. Unfortunately, methane is both colorless and odorless. Also unfortunately, nineteenth-century miners carry fire down into the mines with them, in the form of lamps to see by. The casualties have been climbing over the years, but one explosion exceeds all the rest, killing ninety-two men and boys working in the Felling Colliery near Newcastle.

This is where the hero of the story makes his entrance. Humphry Davy of Cornwall, the father of the field of electrochemistry, was already a scientific celebrity by 1812, having personally discovered or isolated half a dozen new elements. "The intellectuals, socialites and aristocrats of London flocked in droves in the early decades of the nineteenth century to his lectures at the Royal Institution," reads one account. "Such was the congestion caused by the numerous carriages that entered Albemarle Street in the heart of Mayfair (less than a mile from Piccadilly Circus) that it became the first one-way street in the metropolis. Davy's pellucid presentations on science were the talk of the town."[3] This scientist is the great problem-solver approached in 1815 by a trio of men appalled by the Felling catastrophe and the lack of progress since on the imperative to prevent it ever happening again.

Davy embraces the challenge and, after a quick fact-finding trip to County Durham, has a flash of insight: a copper mesh screen around the wick of a miner's lamp, enclosing the flame, could sufficiently hinder the process of ignition to avert a blast (figure 2-1). Meanwhile, just enough methane-tinged air could still reach the wick to make the flame taller, turning the lamp into a methane detector in the same way that a swooning canary could gauge a toxic

FIGURE 2-1

Humphrey Davy's first safety lamp, 1815

Source: Davy's first safety lantern, 1815, and early wire gauge safety lamp, Science Museum, South Kensington, Acme Newspictures, c. December 1929, from the Digital Commonwealth

level of carbon monoxide. Davy sets straight to work on a rigorous series of lab experiments to discover the optimum dimensions of the parts to make it work.

And boom—or rather, no boom—the problem is solved. Just two weeks after he began working on the problem, Davy's safety lamp is fully designed and ready to be produced. Soon it proves itself in the mines, and Davy is honored with the thanks of grateful nations in the forms of honorifics, gifts, and cash prizes. As an older geochemist, John Meurig Thomas, would recount in 2015, "For countless children who, like myself, belonged to the family of colliers, Davy was the first scientist's name that they encountered."[4] His famous lamp remained in use, saving lives, for the next 130 years. End of story.

. . .

The story of the like button recounted in chapter 1 seems a lot more akin to James Hong's descriptions of incremental problem-solving and unclear paternity than it does to the tidy, heroic tale of the safety lamp. In retrospect, the button solved something big by creating the easy feedback mechanism that would provide the mountains of data for untold algorithms to mine for their many purposes. And yet, that was a problem that nobody had set out to solve. How could a feature that turned out to be something humanity now clicks billions of times a day look so inconsequential at the outset? When we spoke with James Hong, we asked whether he was conscious in any way that his work was a step in the evolution of something bigger— ultimately the button that is now the most significant UI piece in the world. His response was a laugh. "When you make things to solve problems, you're not really focused on that," he said. "You focus on what you're making that thing for. I mean, lots of things are invented all the time, and you don't know what's going to take off in the long run in the world and what will not."

There is no single hero of the like button story. Rather, there are lots of unsung contributors who played roles. It was a social process, not a solitary one. In retrospect, at least five people have claimed, or allowed the claim, that they invented it, but unlike the common practice when a new baby is born, no one seems to have sent out proud announcement cards at the time. Like Bob with the sketch he made at Yelp one day in 2005, they all seem to have regarded whatever design or decision they made as really no big deal.

As Jeremy Stoppelman told us, "I'm not a historian and I haven't done all of the research into who did what when—but typically, when there is something new created, there's like a cocktail of different contributors, people working on similar things, all of them adding a little seasoning and salt and pepper from different directions." Pressed a little further, he allowed that prior to Yelp's real-time feedback buttons, he hadn't encountered anything similar on the internet. He also recalled having a dinner with a key early team member at Facebook just a month or two after launching the new feature and freely talking about the success it was having. And surely he spoke to others, too. "In Silicon Valley, everybody shares some of their secret sauce. Maybe not *everything*—but I remember talking about those feedback buttons as well as our compliment system, and so forth. And it wasn't long after that the like button showed up." (In fact, it was a few years before FriendFeed, in 2007, launched its all-purpose, one-emotion option to press a thumbs-up icon.)

Nor did the solution come to anyone in a flash of insight. There was not even consensus on it being a worthwhile thing to do. Mark Zuckerberg, as we have already related, stood dead against it for two full years before Facebook launched the like button in 2009. Soleio Cuervo, the designer who actually drew Facebook's eventual thumbs-up icon in Photoshop, recalled that there was "a lot of concern internally that 'liking' was going to cannibalize engagement."[5] The same concern worried Stoppelman a bit at Yelp. After a prototype of the one-click feedback mechanism had been mocked up, he said, "I remember us discussing it and I remember feeling like, 'Oh, it's sort of in conflict with our compliment system,' because now we'd have *two* ways to get feedback. Should we have overlapping feedback systems?"

But at the same time, there was openness to others' views. Stoppelman recalled expressing these reservations to his CTO, Russ Simmons, but "at the same time, I didn't want to discourage him. He was very excited about this functionality and he felt very strongly. . . . And you know, obviously, we're making stuff up as we go along. You can't have opinions that are too strong about anything. And so, since we had built this working prototype, and it looked interesting, we were kind of curious to see how the community would take to it, and, you know, we kind of moved forward." It didn't bother him that he was overruled by others, especially because the decision was what was known in Silicon Valley vernacular as a two-way door—"where, obviously, you could roll it out and, because it's internet software, you can change things very quickly." If things didn't go as hoped, you could just retreat back through the door you entered by. "And so we rolled it out."

Different things were tried for different purposes. There was definitely cross-pollination happening, as different startups began working with the like button toolkit to solve various problems. Yelp's original problem was, How do we get users generating lots of content without having to pay them? For others, it was crowdsourced quality control. To listen to Facebook's like button team leader, Leah Pearlman, the problem that led to that platform's like button was a need to declutter the profile page, because it quickly filled up with perfunctory reactions that took up lots of space but added very little content. "I felt like I was cleaning the garage by adding the like button," she later told *Fast Company*. "I never expected what happened. . . . We were working on other things at the time that felt like a bigger deal by far."[6] When Instagram decided in 2022 to add a like button to users' stories, it was solving a version of the same problem: cleaning up its direct messages (DMs).[7]

The fact that the like button has been deployed to solve different problems—including problems of its own creation in use—is part of why we saw it start as a multichoice feature, offering a range of emotional responses, then become a single all-purpose positive click, and then become multichoice again. Different variations were launched to varying levels of success. When Google, for example, launched its Google+ social network in June 2011, it refashioned the like button as a "+1," emphasizing the role of the button's counter mechanism and the impulse to add one's bit of affirmation to a growing pile of praise. Selection forces were brutal given the clarity of the feedback mechanism. Things that worked radiated across the social media realm, and things that didn't fell away.

. . .

All this suggests that there is a different way that technology innovation can happen, not so neat and linear.

Brian Arthur puts it more bluntly: innovation is messy. Arthur is an Irish-born economist who came to Northern California to earn his PhD in operations research around 1970 and never left. He is best known for his insights on network effects—that is, the property whereby the value of some products and services rises with the number of customers using them. This is a departure from the typical commercial entity whose operations, past a certain point, are subject to diminishing returns to scale. Networked systems like telephone utilities or package delivery companies—or social media sites—reap *increasing* returns as they grow. The more people participating in them, the more benefits everyone gains from their participation.[8]

But living in Silicon Valley, with a ringside seat to the advent of personal computing and wave after wave of internet-based commerce, has also turned Arthur into a student of innovation. After watching these developments unfold for decades, he developed a theory of innovation that accounts for both a high level of randomness and serendipity in day-to-day problem spotting and solving but, at a higher level, some clear patterns in how technological progress is made.

Arthur observed that as changing conditions give rise to new problems, innovators get to work solving those using their current bag of tricks. And when they come up with an effective solution, that itself then goes into the bag of tricks. So the first thing to be said about how innovation works is that it starts with working with elements that already exist to solve a problem of immediate concern. The social anthropologist Claude Lévi-Strauss, writing not about technology but about how art is created, calls this process bricolage: his neologism draws on the French terms *bricoler* (to putter about) and *bricoleur*, a jack-of-all-trades who has enough facility in different realms to see how pieces from them could constructively be combined. In the same way that a tinkerer reaches for stuff at hand to solve a technical or mechanical problem, Lévi-Strauss says, an artist "shapes the beautiful and useful out of the dump heap of human life."[9]

Arthur's word for this creative process, as a student of complex adaptive systems, is *combinatorial*. New technology doesn't come out of nowhere; no inventions appear out of whole cloth. Instead, new technology solutions are inspired and enabled by elements of past ones and emerge through a process of combinatorial evolution. In his words, "technologies come out of other technologies. If you take any individual technology, say, like, a computer in the 1940s, it was made possible by having vacuum tubes, by having relay systems, by having very primitive memory systems, maybe mercury delay tubes. All of those things existed already."[10]

A fun illustration of how important the base of prior technology is to the attempt to do anything valuable was Thomas Thwaites's toaster project, in which he set out to build an electric toaster from scratch.[11] This meant taking things literally from the ground up by first extracting ore from the earth, then processing it and pulling wire capable of conducting electricity. Even with access to all the inherited knowledge of how to make each component, personally producing them all at a functioning level took him years. Finally the day came that his small kitchen appliance worked! But it performed for only a few seconds before going on the fritz—more than proving the point that any successful technology builder today must depend on the efforts of legions of prior ones.

In his classic article "The Structure of Invention," Arthur calls the preexisting components of a solution "functionalities."[12] But in conversation he tends to use the shorthand of Lego blocks. He told us that people create novel technologies not by "discovering something new, or inventing, but by putting together different Lego blocks, so to speak, in a new way. Once something was put together—like, say, a radio circuit for transmitting radio waves—it could be thrown back in the Lego set." And thus there is relentless progress, a kind of technological evolution—although it is not of the Darwinian kind that operates by natural selection. Instead, Arthur said, new things "come along as completely new combinations, using new principles, and that keeps adding to your Lego set."

One of Arthur's favorite examples is the jet engine, the main assembly of which is made up of five quite elaborate subsystems: intake, compressor, combustor, turbine, and exhaust nozzle. And beyond that central assembly, even more is needed to make the jet work: a fuel supply system, compressor anti-stall system, turbine blade cooling system, engine instrument system, electrical system, and more. It's an arrangement of connected building blocks, each of which is also compiled of preexisting smaller blocks, all together constituting a new combination, which may turn out to have superior or novel functionality.

The most effective innovators are the recombiners who know their Lego sets very well—with a hands-on sense of familiarity with existing technologies beyond what can be learned from reading about them. They can see the latent capabilities. This was a quality that inventor Dean Kamen, best known for his invention of the Segway, was known for. "I wouldn't call Dean an engineer," one of his colleagues once observed. "I'd call him an explorer of the natural world. He's amazing in his ability to pull things back to first principles. . . . Dean understands stuff, just, like, in his fiber. And so he can take that fundamental understanding of how the world works and then take it five levels up."[13]

With this kind of recombination being done by different people in different corners of problem-solving, Arthur notes another common phenomenon. Innovation often displays a convergent property.

The term comes from the realm of natural evolution, where similar features sometimes crop up independently in species that don't interact and where these features weren't present in their last common ancestor. Evolutionary biologists' favorite example of convergence is the capacity for flight, which is shared by insects, birds, and some mammals—but not because they inherited it from the same place. In the realm of human invention, convergence occurs when various people trying to solve problems in their different realms arrive at similar solutions. Often, then, this convergent phase yields to a phase of conscious learning and borrowing from each other.

So this is the other part of Arthur's messy but orderly description of innovation. If you step back from the close-in view of the chaos, it all fits into a grander pattern by which foundations are always having new layers added. And these layers enable new waves of utility: new capabilities are introduced and later leveraged, often to solve different problems.

The vast majority of inventions start out as practical fixes to pressing but rather small-scoped problems, and some of these fixes turn out to be essential keys to larger solutions. And thus, big things often result from the work of people focused on small things. It's the way technology advanced in the past and always will. This is the message that Marc Andreessen took away from a recent interview he did with Brian Arthur.[14] On the one hand, Andreessen felt a bit disheartened by Arthur's assertion. "It's an argument against . . . dramatic innovation," Andreesen said. "Let's just say determined innovation. . . . [T]here's really not going to be anybody in the next twenty years who's going to say, 'I wanna build warp drive'—faster than light travel—and therefore they're going to come up with it. Or immortality, or whatever. . . . On the other hand, it's an optimistic argument, because it says the number of combinations of the Lego blocks are combinatorially effectively infinite over time."[15]

We would add, too, that some of those Lego blocks will be the output of projects that do come out of the gate with whiz-bang ambitions—like warp speed or immortality—and fall short of their audacious goals. A hundred years ago, Harry Grindell-Matthews's prototype of a "death ray," for example, was the subject of excited speculation and thousands of breathless news stories. It never made it past its bench-scale prototype. A quarter of a century ago, Dean Kamen's Segway made it further than that but fell dramatically short of its world-changing projections and has by now pretty much glided off into the sunset. Yet for all their disappointments, these efforts produced technological breakthroughs and commercial insights that proved useful elsewhere.

Another aspect of the order to innovation's messiness is that more widely perceived problems create more foment in the early phase. This is an observation Max Levchin shared with us by way of explaining a strategic exercise he repeats every several years: trying to spot what will constitute the next *substrate* for prolific technology innovation. Levchin, an innovator of internet-based offerings, thinks of that realm as having substrates of its own, underlying the froth of business creation that is visible to consumer and financial markets—and he is convinced that being early to spot an emerging substrate is the key to making the most significant investments. PayPal, for example, was a business he and his cofounders built on the substrate of *transactions*, seeing that they would in general become dramatically more efficient with the advent of internet connectivity. Ten years later, by which point that transactions substrate was being avidly exploited by all kinds of startups, another one revealed itself—what Levchin refers to as the *data* substrate. The internet was producing a tsunami of data that wasn't yet being integrated into the world of online transactions. This is the substrate on which he moved early, in 2012, to found a next-generation credit network called Affirm.

Levchin is underscoring that fundamental shifts in underlying conditions—like a new superefficient transaction mechanism or a new superabundance of data—create many opportunities for many innovators to explore. They create puzzles and problems that did not exist before or that were not within the realm of practical reality to solve. And therefore they give rise to an efflorescence of activity at the next level by many smart and energetic problem-solvers, each of whose work inspires, challenges, and augments the efforts of the others.

. . .

Another aspect of innovation seems to be stubbornly ignored or mis-construed in the retelling of the stories. As well as being a much less neat and linear process than it is often portrayed, innovation is far from solitary. And collaborative innovation is certainly more prevalent today than in, say, the eighteenth century, when Humphry Davy and his far-flung scientific peers corresponded by handwrit-ten letters. As described earlier, in the realm of social media and UI development, Saul Griffith and his wife Arwen O'Reilly hosted monthly collaborative meetups at Squib Labs. New developments and ideas were constantly being made visible and actively discussed there, served up rather instantly for others to borrow and build on.

Hong's comments at the outset of this chapter are notable be-cause they speak to the reality that innovation is a distributed phenomenon—across people and across time. Everyone learns from each other, everyone stands on the shoulders of giants, and there are more shoulders to stand on with every passing year. With every pass-ing decade, there is more collaboration with colleagues. As Magda-lena Skipper, editor of one of the world's leading scientific journals, recently told an audience at Davos, Switzerland, "When *Nature* was first published back in 1869, single-author scientific publications were the norm. They are not only an exception but a true rarity today. Science has become a team activity."[16]

The Nobel Prizes reflect this, despite their original heroic orien-tation. When Alfred Nobel wrote his will in 1895, devoting a large chunk of capital to a fund to award great contributors to human progress, he specified how the money should be distributed each year:

> The interest is to be divided into five equal parts and distributed as follows: one part to the person who made the most important discovery or invention in the field of physics; one part to the person who made the most important chemical discovery or improvement; one part to the person who made the most important discovery within the domain of physiology or medicine; one part to the person who, in the field of literature, produced the most outstanding work in an idealistic direction; and one part to the person who has done the most or best to advance fellowship among nations, the abolition or reduction of standing armies, and the establishment and promotion of peace congresses.[17]

In chemistry, for example, the first prize was granted in 1901, and it wasn't until 1929 that two chemists were credited with a discovery and jointly celebrated. Contrast this with the past twenty years, when fully thirteen prizes have gone to teams of three. No single scientist's contribution can be isolated as being the most essential to the advance in question. And in truth, those three names don't even scratch the surface. (The Nobel rules limit the honorees to a maximum of three.) Even with innovations that are universally hailed as important feats of human ingenuity, many of the people who played pivotal roles in their emergence often get little or no credit.

This is not a situation that bothers James Hong in the least. In our conversation, he pointed to a bit of industrial design in a flip phone he had at hand, and said, "Someone thought, like, 'Oh, you know, to make this work better, I could . . .' And they created this hinge. I *appreciate* that I have no idea who did that—honestly, I don't care, just like no one out there should care, every time they click a button on a screen, that 'Jim Young made that for us,' or whatever. Especially because maybe someone else came up with this hinge first—who knows? It doesn't matter. The point is that everyone's building new things and mankind is progressing, and life is getting better and easier for everyone. We're all just making these little micro-contributions."

. . .

Another way that innovation seems to depart from the way people talk about it—beyond its being not so linear and not so solitary—is that there isn't much foresight involved. This was definitely the case with the like button. Nobody at the outset foresaw what, in retrospect, is its most significant result, the aggregation of data about likes to revolutionize the field of marketing and advertising. The initial focus was way down at the granular level of making one person's reaction visible to another individual. In Yelp's case, it was making reviewers feel good about their writing and motivating more content creation. In HOTorNOT's case, it was giving photo uploaders the unvarnished assessment they sought (and possibly sparking a connection). Only further down the road did these granular responses add up to a mountain of data to be mined and exploited.

Innovation unfolds like this all the time. People don't see the ultimate prize at the outset. Sergey Brin and Larry Page's development of PageRank is a modern example. The idea was to go beyond the prevailing way of rating the quality of a page of content, which had simply been to see its number of hits—that is, how many visits had been made to it. Taking a lesson from the world of scholarly citations, their insight was to also assess the importance of a page of content in terms of the number and quality of other pages that had *linked* to it. A page that is linked to many pages with high PageRank receives a high rank itself. Brin and Page knew it would be important to devise a better way to rank page quality, because marketers were starting to run banner ads on pages. But Page Rank really paid off by providing the basis for a superior search engine. Brin and Page weren't envisioning Google's eventual business model as they focused on page ranking, but then, after they landed on a reliable algorithm combining hits and links, little else had to be done to launch Google. (Very rarely can innovators predict, either, what new problems will crop up to solve once they've unleashed some new thing on the world and it begins to be taken up. And they don't see the new problems their innovation will cause. It's a topic we'll take up in chapter 7.)

Innovations are also serendipitous in the sense that none of them is inevitable. Back in 1997, when Andrew Weinreich launched the world's first social networking site, Six Degrees, it wasn't obvious to him that the site should have a like button. Nor was it obvious as Friendster, LinkedIn, and Myspace arrived on the scene, or to any other site designers for years to come. It was fully a decade later that Facebook decided a like button would be a good thing to add, and only after Zuckerberg had rejected the idea for two years. Had this one powerful person stuck to his guns and deprived the like button of the 2009 deployment that caused it to go exponential, would the feature ever have become as big as it did?

Even when all the elements exist to make a new invention possible, and when a crying need for a better solution would make it undeniably valuable, it can be derailed by any number of things. This is a truth brilliantly illustrated by the classic Alec Guinness film *The Man in the White Suit*. It's the comic story of a scientist working in a business R&D department in Northern England who comes up with a miracle textile—made of fiber akin to polyester and perfectly comfortable but impervious to stains and able to last forever. What a boon to humanity, he recognizes: one might only need buy a set of clothes once and be free of that expense forever. Soon, however, he finds he has a target on his back as it becomes obvious how antithetical that is to the interests of not only the mill owners but also the labor unions—and beyond them, the haberdashers and dress shops, launderers and tailors, and every other business predicated on the current imperfect realities of fabric. There is no way the system around him will let this innovation come to market. Much conniving and thwarting, and a memorable mob chase scene, go into the effort to keep him from succeeding.

The point is that any innovation is embedded in and must succeed in the context of a social system. As well as the fortuitous conversations and learnings that spark the idea, it must wend its way through regulatory frameworks, the structures of industry sectors, the transportation infrastructure, the capital markets, the values people live by, and more. And because big changes require more than awesome tech advances, high-impact innovation usually starts small and grows out of unassuming solutions. Indeed, something that comes out of the gate looking like it will be very big in its implications may only suffer from that promise of greatness by calling too much attention to itself and drawing fire from whatever realm it threatens.

.   .   .

So perhaps the past great inventions whose stories we know and love best are actually anomalies, not reflective of how innovation usually works. Or maybe the dramatic difference is really a matter of how the tales are told. We're increasingly inclined to think the latter is true because, with our new mental model of innovation as a messy, social, and myopic process, we decided to dig deeper into some historical breakthroughs. In each case, we found far more complex dynamics than you would suspect from the celebrated accounts.

Even the heroic tale of the safety lamp turns out to be nuanced, starting with the fact that Davy was hardly the first to work on the idea. Up north in 1815, the same year that Davy was working on the problem, George Stephenson, an engine-wright associated with collieries in the Killingworth area, invented what he called the Geordie lamp—not to echo his own given name but to honor the Geordies, the folks of Newcastle and the surrounding area of Tyneside. And fully three years earlier, a local doctor named William Reid Clanny had also come up with not just the idea of a safety lamp but a working model.

Different variations were designed and tested. Clanny's version had a large candle surrounded by a glass casing and, at its bottom, a pool of water through which air would be purified before reaching the flame. It worked, but at the cost of an extra boy to wheel around the contraption and keep it alight by working its bellows to push air through the water. Clanny kept at the task, introducing a different design in 1816. Davy, too, after producing the lamp that was put into wide service, continued to refine its design. Learning that it failed, for example, in especially drafty conditions, he added extra shields and gauze cylinders to stand up to gusts of air. Incidentally, Martin experienced this limitation firsthand, when using Humphy Davy's celebrated lamp design in a demonstration as part of a lecture on innovation and imagination at the Royal Society. The lamp provided no protection from the draft from the butane aerosol and a small explosion was triggered.[18]

And all this was done in a social context that encouraged collegial knowledge-sharing. Clanny, who made his living as a physician, was purely motivated by the appalling injuries he treated in the miner population. He was a founding member of the Society for the Prevention of Accidents in Coal Mines, established in 1814. Down in London, the Royal Institute was the setting where many inventors attended meetings and presented their latest work, and in 1812, the institute featured Clanny presenting his paper "On the Means of Procuring a Steady Light in Coal Mines Without the Danger of Explosion."

We might also hold up the example of the steam engine. Do you know what its first *few decades* of practical applications focused on? If you answered, "Pumping water out of flooded mines," we're impressed—and we're happy to award extra points if you also summoned up a launch date of 1698. To most people, that would sound way too early as well as a rather inauspicious start for the invention now broadly credited with ushering in the Industrial Revolution through its disruptive impact on transportation, manufacturing, mining, and agriculture. The journey to that outcome was not straightforward and efficient. It involved a social process of many tinkerers making progress in fits and starts. There was no one aha moment of genius insight or visionary foresight.

Pretty much the same is true of every invented thing you see around you, including all those things you don't think of as ever being invented at all. Yet the messiness of innovation is often obscured in the later telling of the story. As in all storytelling, extraneous details get edited out and the plot focuses efficiently on the sequence of essential steps from novel idea to nailed-it implementation. The cast of characters is also winnowed down, and the tale ends up conforming to what Joseph Campbell called the hero's journey: we hear how a single protagonist we learn to care about embarked on a purposeful quest and made clear-eyed progress through various challenges toward an achievement that ultimately benefited humanity.

Why do the stories told about innovation not relate the process more honestly? Maybe the point of telling them in the first place is to motivate individual initiative. That is not an unimportant goal, of course, because personal drive does make a difference. Take the story of Craig Venter, his development of shotgun genetic sequencing, and his conviction that it was a better technique for decoding the human genome. You do really have to hand it to him: he was right, his work was paradigm-changing, and his name deserves to be celebrated. But to earn his place in history, he had to burn a lot of capital—both financial and social—because, just as in the Alec Guinness movie, there was a large establishment invested in the status quo.[19]

That innovations are often thwarted by such establishments isn't just the stuff of movies or a modern result of supersized bureaucracies. Here's how a historian of technology, writing in the mid-nineteenth century, described the same phenomenon:

> In looking over the history of great inventions it is remarkable how uniformly those discoveries that helped mankind most have been derided, abused, and opposed by the very

classes which in the end they were destined to bless. Nearly every great invention has had literally to be forced into popular acceptance. The bowmen of the Middle Ages resisted the introduction of the musket; the sedan-chair carriers would not allow hackney carriages to be used; the stagecoach lines attempted by all possible devices to block the advance of the railway. When, in 1707, Dr. Papin showed his first rude conception of a steamboat, it was seized by the boatmen, who feared that it would deprive them of a living. Kay was mobbed in Lancashire when he tried to introduce his fly-shuttle; Hargreaves had his spinning-frame destroyed by a Blackburn mob; Crampton had to hide his spinning-mule in a lumber-room for fear of a similar fate; Arkwright, the inventor of the spinning-frame, was denounced as the enemy of the working-classes and his mill destroyed; Jacquard narrowly escaped being thrown into the river Rhone by a crowd of furious weavers when his new loom was first put into operation; Cartwright had to abandon his power-loom for years because of the bitter animosity of the weavers toward it. Riots were organized in Nottingham against the use of the stocking-loom.[20]

The determined opposition painted here suggests a defensible reason that the lone inventor is so mythologized in the stories we tell about innovation. If even the most valuable new technologies have to be "forced into popular acceptance," then human progress depends on the occasional iconoclastic thinker who is determined enough to do that forcing—and inspirational stories can be part of building that determination.

Research by neuroscientists suggests that the human love of a good character-driven story is tightly connected to our evolutionary psychology and particularly our penchant for social learning.[21] Stories are the most memorable form for sharing information. They model the world around us and teach problem-solving. Telling heroic tales might encourage young people that they can make a difference (and should maybe study harder)—and fortify them better against the slings and arrows that will come their way if they come up with anything that meaningfully challenges conventional wisdom.

. . .

For an inspiring tale from Silicon Valley, you could do worse than the one we heard from Max Levchin when we asked him about the Gausebeck-Levchin test mentioned by James Hong. It was a solution developed at PayPal, and it soon spread across the web like wild-fire, ultimately to be used billions if not trillions of times. But sure enough, Levchin told us, "The original story of our invention of it was pretty simple." He said it focused narrowly on a very specific, pressing problem for his team.

On a daily basis, PayPal was being inundated with new account openings by hackers who were creating them to swamp the company's ability to do the know-your-client compliance work demanded of all players in the financial services industry. Exploiting the online system of account application and approval, the hackers were using automated scripts to pose as real customers and open thousands of accounts every day, creating an unmanageable environment in which illegal money-transferring activity could slip by undetected. This devious scheme represented an existential threat to PayPal's ability to keep doing business. "What I needed to do was reduce the number of fraudulently created accounts," Levchin explained.

Levchin's team knew from the start how the solution would have to work. "The goal," he said, "was to make it impossible to create these new unauthorized accounts without at least some amount of human effort." But every time they figured out some sort of hack to slow down the onslaught (e.g., by making the names of the html form fields unpredictable), it was quickly overcome. "Two days later, the bad guys would figure out how to get around it—because anything that doesn't require actual human work to do can be automated, and that's what they did." To this day, Levchin remembers with absolute clarity when the breakthrough came. "I walked out onto the engineering part of the floor and sort of screamed to the various engineers nearby, 'What is one task that humans are really good at and computers are really bad at?' And this guy Dave looked up and said, 'Optical character recognition.' Which was, *yeah!*"

Adding relish to the story is that the hackers were not only crafty but also cocky. One in particular, whom Levchin pegged as a fellow Eastern Bloc type since he misspelled his alleged name as Greg Stivinson, liked to send taunting emails after finding a way around the latest PayPal attempt to thwart him. It was the closest experience, Levchin now recalls laughing, that he ever had to a *Hunt for Red October*–style mind game with an arch nemesis. But back then, Levchin thought he might have his kill shot.

His first step was to jump in his car and head down the street, because "optical character recognition was one of these things you could buy a piece of software for. You could scan a handwritten sheet and upload it into your OCR software, and it would spit out a PDF file—or a text file converting the visually recognizable characters into type." But even when handwriting was readable to other people, it was often impossible for a computer to make sense of the scribble. "So I drove to Fry's Electronics, which was nearby the PayPal office, and bought every copy of OCR software that money could buy. I literally had a stack of these boxes—like, eight different packages." Then he wrote a quick piece of code to generate a quickly messed-up piece of text and to ask the user to retype it. "And the way I knew it would work is, I ran it through all these OCR packages, and there were ways that they all failed.

"So I coded from Friday night till Monday morning," Levchin said, "with no sleep, no anything. And this was one of the very last times I personally deployed software to PayPal's website myself, from 'How do I solve this problem?' to just being so delirious that I just pushed the code to life." It must have been a euphoric experience because he recalled that as he saw the first of his coworkers arriving at the office

Monday morning, he climbed up to a sturdy corner of his cubicle walls and assumed a lotus position . . . and later, once everyone had arrived, he blasted out the "Ride of the Valkyries" through speakers he had originally bought for his college dorm.

And the crowning touch was that when Levchin inserted that particular chunk of code—the piece he had just "built separately in a total sprint over two days of insane programming"—he titled it with the term *stiv*, in honor of that guy who had been goading him. Better yet, the guy saw it and, a few days later, sent a two-word email we won't directly quote here. "Stivinson" never opened another account, and the problem was solved.

All in all, it's a great story with some classic elements: a mortal threat demanding a creative solution, a protagonist willing to venture forth, an archetypal underworld villain to be dispatched, setbacks and suspense, and ultimately a victory to be hailed (complete with Wagnerian score). In the hands of a master storyteller, could the tale be made even more engaging? Well, sure. We could start with the obvious: Where's the love interest? But probably the other problematic element that would send the script back into rewrite is the epilogue, where we learn that this isn't *the* story behind the captcha tool we all know. It's just *one* story behind captcha.

Again, captcha was born in at least two places—and went on to grow fast as many other site builders saw the huge potential of a "reverse Turing test" to authenticate online visitors. Why didn't Levchin and his team see the future of the feature, get more excited about it, and lay claim to a bigger prize? In a word, they were swamped. "For context," Levchin said, "this was just after the Elon ouster, and we were still losing money hand over fist, and there was horrible fraud everywhere. This was just one of the many whac-a-mole things that we had to do to get it right. So it was definitely not a thing we were considering commercializing or thinking about through the lens of, like, 'Oh, what an interesting or important invention.' It was just survive or die." It's an all-too-relatable, if somewhat deflating, denouement.

But maybe this is the way we need to start relating the histories of important innovations—as collections of stories, each individually engaging but incomplete, combining to get at the truth. It might be too much like the story of the blind people describing an elephant—a tad too kaleidoscopic to take in as a satisfying narrative arc. But because of those very qualities, people would go much further to portray the process of impactful innovation in commercial ecosystems everywhere the way it actually works.

. . .

If that's how innovation works, in fits and starts from many directions by many people, what does it imply for those who are trying to get more of it? We know that all kinds of organizations, along with the economic development offices of many locales, are trying to innovate. How and what should they change to encourage more inventiveness?

One thing we should learn from the like button and other warts and all histories of innovation is to embrace an unruly process, not pretend it is otherwise. This important lesson flies in the face of how many of us have learned to think about commercial processes in general. In a textbook framework, innovation can be made to seem like a purposeful, reliable, and even efficient organizational activity. It can look like a very manageable task of finding novel solutions and translating those visions into reality through sustained work.

But because the reality is more chaotic, more myopic, and more serendipitous than this linear view implies, we need to manage with more of a mindset of cultivation than tight control and one that may be characterized by more setbacks and surprises than intended outcomes. Large companies, in political scientist James March's memorable phrasing, are set up for *exploitation and not exploration*. They are good at taking full advantage of past breakthroughs, efficiently cranking out products and services that existing technologies, materials, and recipes make possible, improving the efficiency of their production, and making those wonders available to larger swaths of society. When they turn their hands to exploration—trying to discover the new new thing—their approach by long habit is to try to structure and streamline the innovation process. But their attempts to de-risk it and make it more efficient, predictable, and straightforward ignore the basic facts of a process that is inherently messy and serendipitous.

We need to be humbler about what we don't know and can't predict. We can't talk about innovation in Silicon Valley software without mentioning the sheer audacity that has taken hold in recent years—for example, Google's moonshot division, the X Prize crowd, and the general ambition to solve fundamental problems.

Part of this humility is acknowledging that technology breakthroughs don't exist in a vacuum. We mentioned Saul Griffith in passing above—the guy who, with this wife Arwen O'Reilly, created Squib Labs. When the *New Yorker* profiled Griffith, a MacArthur "genius grant" recipient and a serial inventor, writer David Owen told the story of his early work to manufacture eyeglasses less expensively to make them available to poor people around the world. Ultimately, Griffith realized he was tackling the wrong problem. It wasn't the cost of the lenses that was limiting access—the real impediments were in the distribution and service infrastructure. Owen wrote that since then, "his thinking has been deeply influenced by what might be thought of as the cheap-glasses conundrum: the inadequacy of addressing complex societal issues with technological ingenuity alone. Nowhere is this problem more apparent than with his main preoccupation these days: energy use and global warming. The world's most urgent environmental need, he has come to believe, is not for some miraculous scientific breakthrough but for a vast, unprecedented transformation of human behavior. That conviction makes him doubly unusual: an extraordinarily innovative engineer who is trying to think his way around the limits of innovation."[22]

Brian Arthur has a phrase to describe the frustration Griffith must have felt. It often takes decades for technology domains—that is, collections of technologies that use similar phenomena or are functionally similar—to penetrate markets. This is because, in Arthur's words, economies encounter new technology domains. These encounters typically require substantial change in the economy, including changes in financing for new technology domains, the creation of new jobs and associated training regimens, and even the redesign of the physical infrastructure and organizational structures and processes. For example, one of the reasons that it took decades for electric motors to replace steam systems in US factories was that the factories themselves needed to be redesigned to take advantage of them.

We need to focus on experimentation, feedback, and an iterative process. Maybe the biggest mistake made by big, long-established corporations attempting to manage innovation processes is that they read and believe the accounts of famous inventions that portray the process as purposeful, linear, and disciplined. Such accounts are an encouragement to try to put the process on rails—and to make one powerful person responsible for it. A typical approach in large firms is to appoint a chief innovation officer. The officer is typically given a spot in headquarters and a dedicated staff, and their role is often to map out and then impose processes by which new ideas will be surfaced and vetted for their commercial potential. These ideas' early-stage development will then be conducted in a standardized fashion, in an environment appropriately protected from the performance pressures of the company's mature businesses.

This is a challenge that Walmart has tried to tackle, for example, with its insistence on overcoming the tendency of large organizations to create many layers of managers, any one of whom can kill an innovative idea with their no. Instead, the retailer created the processes to clearly define problems and to inculcate an institutional bias toward saying yes, at least to testing the proposed solutions. This approach has likely been one of the keys to the steep rise in Walmart's share price in recent years. Tom Ward, chief e-commerce officer, told us that when creating the capability that allows customers to order online and pick up in-store, for example, the company first launched an app that required the user to be identified on arrival. This app was followed by a version that allowed manual check-in via an app. Followed by an automatic check-in via geolocation. "If you're improving on a solution people love," Ward said, "you're probably only going to run into people that green-light the road. But if you're trying to build something that's perfect, you're going to have ten different versions of what people think perfect means and you're not going to deploy anything. We focus on the twenty that's worth the eighty, get something out there, and iterate." Interestingly, today even companies with two million employees can sound a lot like a hundred-person startup.

To be sure, not every realm of human endeavor can emulate the innovation processes of Silicon Valley. As Owen puts it in the *New Yorker*, "The speed with which software-based activities and Web innovations catch on—text messaging, eBay, Twitter—has encouraged a public perception that transformative technological change takes place almost instantaneously." Change is harder to make when you're dealing with atoms, not just bits. Still, many old-dog companies could do a lot more if they got serious about learning some new tricks.

. . .

Here's what we set out to answer in this chapter: What does the like button have to teach about how innovation works? And we conclude that although innovation generally works in the messy fashion epitomized by the like button, on a higher level there are clear patterns. Between the nature of its products—digital—and the conditions for social learning, Silicon Valley became a hotbed for this type of innovation. All those factors that tend to get cleaned up too thoroughly by efficiency-minded mature corporations were left in place to do their magic.

The story of the like button serves well as a case study for this messier model for innovation. Mainly we found a corrective to the romanticized stories told about the innovations of the past. They aren't helpful in today's context, and they aren't even truthful about the past. We saw that it's really the norm, throughout history and today, that innovation is nonlinear, combinatorial, social, and unpredictable. And we also saw that these things are all especially true when we're talking about digital innovation.

Invention is not how it sounds in the retelling. The story always sounds like a straightforward and brilliant march from problem to inspiration to perspiration to solution. But the reality is more like the like button. The creation of the button is not just an interesting story of a wildly successful innovation, but also Exhibit A of how software innovation worked in Silicon Valley at the turn of the twenty-first century and how it increasingly works everywhere else.

In chapter 8, where we look at the future of the like button, we'll assume the same kind of process is driving its further evolution. It will be messy, and it will produce new value along with unanticipated side effects, all of which will look inevitable in hindsight but will still manage to surprise us. As further evidence of the circuitous path of many innovations, the next chapter describes the surprising journey the thumb symbol took as it hitchhiked its way through American history.

Chapter 3

Why the Thumb?

To visit the district of La Défense in Paris and stroll through the square known as Place Carpeaux is to be struck by an arresting sight: there you encounter a forty-foot statue of a human thumb, pointing proudly skyward and weighing in at some eighty tons of bronze (figure 3-1). For its creator, César Baldaccini, *Le Pouce* was a highly accurate "expansion" of his own digit (down to the fingerprint), originally rendered in 1964 in that wonder material of the era, polyurethane, at a rubbery pink height of seventeen inches. Later, César would scale it further and in other media, captivated by the work he had initially described as just "part of a goal I have assigned myself—I want to make my portrait with fragments of my portrait."[1] But whether or not you knew that this choice of body part was meant as a starting point or that the work had been inspired by casts being made by archaeologists of ancient victims at Pompeii—or really any of its backstory—you could be forgiven for asking as you came round the corner from the train station: Why the thumb?

FIGURE 3-1

César Baldaccini's *Le Pouce* sculpture

Source: imageBROKER.com GmbH & Co. KG / Alamy Stock Photo

It's the same question Martin asked when Bob showed him his early-concept sketch of a like button. Why had he drawn it with a thumbs-up? And why did other interface designers gravitate to that icon as well? Was there something particularly resonant about it? Did they consider it especially friction-free? As described earlier in the book, the like button's hockey-stick takeoff in usage came only later, after Facebook had launched it—with a thumbs-up symbol. And sure, that uptake was a matter of Facebook's massive user base, but was it also something about the thumb?

It turns out that, yes, there were strong reasons that the thumb would come to mind and that incorporating it would boost the uptake of the like button. A well-beaten pop cultural path had promoted the thumbs-up image, refined it, and reinforced it across much of history. By lifting an iconic in-real-life (IRL) gesture that everyone knew and many used, designers at the time were repurposing an already-established behavior and the appropriate positive sentiments it could express.

But there are even deeper reasons why the thumbs-up became such a common IRL gesture in the first place. So as we tried to trace the origins of a simple UI design idea conjured up in 2003, this turned out to be another rich story and another fascinating journey of discovery.

. . .

"Are you not entertained? Are you not entertained? Is this not why you are here?"[2] Probably you recognize the quote, and if so, you are picturing the scene: Maximus, played by Russell Crowe in the 2000 blockbuster *Gladiator*, is shouting his rebuke to a Roman amphitheater crowd that had earlier been baying for blood but has just been shocked into silence by his over-the-top savagery in the arena. In the current social media environment, the clip gets constantly shared and reposted, a *Gladiator* meme second only to the film's most potent image: Joaquin Phoenix as the emperor Commodus extending his right hand to issue a verdict of thumbs-up—or, more fatefully, thumbs-down (figure 3-2).

When Yelp and many other social web startups were launched, Ridley Scott's epic *Gladiator* had only recently come out. Streaming was not yet a thing—Netflix was still physically shipping DVDs to its three hundred thousand or so subscribers through the postal service—but between the film's spectacular box office debut and Oscars-fueled VHS and DVD sales, seemingly everyone had seen it.

A decade before *Gladiator*, the release of another blockbuster, *Terminator 2: Judgment Day*, had been another banner day for the thumbs-up. Its famous final scene sees the hero sacrificing itself in a vat of molten steel but managing to keep one hand aloft—and in its last moment, signaling a thumbs-up to the human race that it has

FIGURE 3-2

Thumbs-up from the movie *Gladiator* (2000)

Source: Landmark Media / Alamy Stock Photo

saved (figure 3-3). Five sequels later, Tim Miller, director of *Termi-nator: Dark Fate*, would admit he wasn't a fan of that ending: "Jim Cameron loves that scene, many people love that scene. It's . . . iconic, I know, but maybe a little too sentimental."[3] Still, he recognized the impact it had on its core audience: "I can only say that that is not my favorite part of the movie, and the fact that it is yours is a symptom of your age when you saw it. Because when I saw *T2* I was, like, twenty-seven." Back in 1991, many of the designers working in Sili-con Valley during the early 2000s had been in the adolescent sweet spot to better appreciate that scene.

Besides being an affirmative hand gesture, the thumbs-up had also by 2003 been very much embedded in American culture as a rating system, thanks to film critics Gene Siskel and Roger Ebert and their television show *Sneak Previews*. To avoid an intellectual property dispute in 1986, when they took the show from public broadcasting to network distribution, they had to make some format changes. In the original PBS show, Ebert later recalled, "we had yes

FIGURE 3-3

Thumbs-up from the movie *Terminator 2: Judgment Day* (1991)

Source: TCBS Photo Archive / Getty Images

votes and no votes and . . . I suggested, why don't we just say 'thumbs-up' or 'thumbs-down' instead?"[4] Soon enough, whenever both the argumentative hosts agreed a film was good, that studio's marketers would seize on their "two thumbs-up" as high praise. "[The two thumbs-up] would appear in a lot of movie ads and it became a famous catchphrase," Ebert said. This turned out to be a "big deal," he said, since "the concept of something getting two thumbs-up did not exist in the English language until we did it on the show." When we recently sat down with Phil Libin (former CEO of Evernote, born in Leningrad, USSR [today's Saint Petersburg], and one of Silicon Valley's most well-regarded product people), he shared the perspective that "the thumbs-up icon could only have come from America. Looking back, it was the perfect icon for that place and time."

So on one level, any UI designer favoring a thumbs-up icon was just reflexively reusing a symbol that had been hammered into the public consciousness around the turn of the millennium—but on another level, those uses of it had been derivative, too. Google Ngram

Viewer, which charts trends in word and phrase usage over time, shows that "thumbs-up" had really started taking off in the mid-1970s. What would have fueled its rise then? The prime suspect has to be the decade-long, top-rated TV comedy series *Happy Days*, which aired from 1974 to 1984. Its most popular character, played by Henry Winkler, was "the Fonz" (more precisely, Arthur Fonzarelli), the one tough guy from the other side of the tracks inserted into the show's squeaky-clean 1950s social scene. The Fonz's signature move, usually accompanied by a super confident "Ayyyy," was a double thumbs-up (figure 3-4).[5]

Perhaps it's hard to imagine the likes of Ridley Scott, James Cameron, and Roger Ebert, even at a more tender age, taking much inspiration from the Fonz. But note who Winkler was rather obviously channeling with his performance. This was a moment when pop culture was dramatically advancing the mythology of James Dean as a tragic figure. A symbol of restless disaffection, given his starring role in *Rebel Without a Cause*, Dean had also been an avid motorcyclist and race car driver and met his end very young in a high-speed highway accident. In a famously haunting photograph taken mere days before his death, he sits in his Porsche Spyder and greets the camera with—you guessed it—a thumbs-up (figure 3-5).

It would have been a habitual gesture for him, given his favorite recreations. Motorcyclists used it and a variety of other hand signals to communicate with fellow riders nearby over the din of their unmuffled engines. Others, in turn, found it irresistible to mimic their edgy, outsider culture. When Dean proudly brought home his Triumph TR5 Trophy motorcycle, he made no secret of its being an homage to his own personal hero: Marlon Brando had ridden a Triumph Thunderbird 6T in the 1953 hit *The Wild One*.[6]

But we can keep unspooling this thread, because in using such gestures, the bikers of the 1950s were themselves quoting the past. They were carrying forward the habits of motorcycle troops in World War II, where the "trusty Triumph" served so many soldiers well. Harley-Davidson's WLA model came to be called "the Liberator,"

FIGURE 3-4

Henry Winkler as "the Fonz," giving two thumbs-up

Source: Everett Collection Inc. / Alamy Stock Photo

FIGURE 3-5

**James Dean in his Porsche Spyder, mere days before
his fatal car crash**

Source: Bettmann / Getty Images

because European civilians so strongly associated it with how they saw it being used: escorting convoys.[7] And to take the gesture back even a step further, the motorcycle troops were surely borrowing a habit already established among those dashing heroes of both world wars, the pilots.

In what is perhaps the first print reference to a thumbs-up meaning "all good," a World War I veteran named Arthur Guy Empey includes it in a glossary appended to his memoir *Over the Top: By an American Soldier Who Went*. Empey published the book in 1917, having impulsively decided in 1915, after the sinking of the *Lusitania*, to cross the ocean and enlist in the British army. His gripping account of fighting with the Tommies in the trenches and surviving the "big push"—the Battle of the Somme—made *Over the Top*

America's top-selling book of the year. According to the wry definition in his "Dictionary of the Trenches," *thumbs-up* is "Tommy's
expression which means 'everything is fine with me.' Very seldom
used during an intense bombardment."[8]

That book predates the formation of the Royal Air Force by a
year, but airplanes had already been in military use. For pilots, like
motorcyclists, the habit of using a thumbs-up gesture began out
of necessity: encased in the cockpit of a noisy machine, the pilots
needed a way to signal preparedness for takeoff to crew on the tar-
mac. But the gung ho positivity of the gesture also radiated heroic
bravado. By the time the United States entered World War II, flashing the thumbs-up was signature aviator behavior (figure 3-6). It
was the stuff of endless photojournalism—and no small measure of
propaganda. Throughout the Cold War, the thumbs-up served as a
worldwide symbol of US competence and confidence. A news account shows that in 1955, familiarity with the gesture had reached
the point that the beleaguered Chinese peasants of a tiny island near
Taiwan knew to flash it as they were being evacuated by US troops.[9]

Given all this, is it any wonder that a UI designer in America wondering how to represent a like would think first of a thumbs-up? This
is a story of an icon that had been reinforced in many waves before
the social web came along. It was part of the cultural water the social
web startups were swimming in, whether they saw it consciously or
not. But on the other hand, the story must go deeper, because where
did the World War I soldiers and pilots get the thumbs-up?

. . .

The answer is a uniquely American one, but it also takes us back to
the ancient gladiators. One reason the thumbs-up expression resonated with Americans in the early 1900s was that so many had heard
and read the fiery sermons of the Reverend Thomas De Witt Talmage. Disdained by some in his time as a "moneymaking preacher,"
he crisscrossed the nation, giving paid speeches that tended toward

FIGURE 3-6

Thumbs-up in US war bonds poster

Source: Bettmann / Getty Images

the sensational. Decades after his death in 1902, he was still being declared as "one of the greatest pulpit orators this country has produced and probably no other clergyman in the country's history reached as many people by his preaching as did Dr. Talmage."[10]

The famous preacher's reach had much to do with his knack for storytelling and his charismatic presence, but there was a technological innovation behind it, too. "His sermons were given their peculiarly wide circulation," a 1902 obituary explains, "through the means of plate associations. That is, they were distributed in form ready for press use to thousands of the smaller dailies, which, in addition to their use in the city newspapers, gave to his sermons a circulation much greater than that of any other regular contributions to current literature."[11] The estimate was that his words reached "probably 30,000,000 persons every week"—a number representing about 40 percent of the total US population of the time.

Talmage often worked into his sermons a vivid reference to the verdicts rendered in the Roman amphitheater, as in this 1895 sermon delivered in Washington, DC: "Sometimes the audience came to see a race, sometimes to see gladiators fight each other, until the people, compassionate for the fallen, turned their thumbs up as an appeal that the vanquished be spared, and sometimes the combat was with wild beasts."[12] Or, consider this riff in a sermon on Saint Paul's words in 2 Timothy 11:24, "Be gentle unto all men":

> In the great coliseum of the world's struggles we should never
> do as did the ancient aristocrats, brutalized by the spectacle
> of the gladiatorial combat. When the victor in one of these
> fierce fights had beaten his foe to the earth, he looked up to the
> spectators for a signal to tell him if his prostrate enemy was
> to be killed or to have mercy shown him. Then even women
> so far forgot the tenderness of their sex as to turn down their
> thumbs, which was the signal to finish the wounded man. We
> should always lift our thumbs up. We should not give the signal
> "to destroy." We should signal always "to save," no matter how

much the prostrate foe may hate us and our work. If this Pauline commandment bids young Timothy to be gentle with his friends, it also commands him to deal gently with his enemies.

And again, he preached:

In the old Roman Coliseum, when a gladiator was flung, the angry foe standing over him had first to look up to the vestal virgins to find out whether he might spare him or plunge into his heart the fatal sword. If the vestal virgins wanted the prostrate fighter to live, they would point their thumbs to the heavens. If they wanted him to die, they would point their thumbs down. So today by the lessons of the bargain counter we hold the life and the death of multitudes as they did in the Roman Coliseum. Shall we say "Thumbs-up" or "Thumbs-down"?[13]

Clearly the invocation of the decadent Roman circus could liven up any sermon advising on how to make a moral decision and anticipating that there would be a day of judgment for all sinners. This was a man, then, whose work the public knew very well, who was a powerful vector of cultural influence, and who made heavy use of the thumbs-up/thumbs-down metaphor. Talmage's words, carefully crafted for maximum impact, helped to plant the phrase and image of thumbs-up into the soil of American culture.

As to where Talmage himself picked up his thumb imagery, there is no question. Having gained a reputation as a crowd pleaser in Philadelphia, he had been called in 1869 to lead the Central Presbyterian Church in Brooklyn, New York. There he set up a household full of art and artifacts. As a New Yorker, a collector of paintings, and a history buff, he certainly would have seen and pondered the Jean-Léon Gérôme painting *Pollice Verso* (figure 3-7).

The vivid historical painting was first exhibited in 1873 at a private salon in Paris but was soon purchased by a New York retail magnate, Alexander Turney Stewart, and shown to American art

FIGURE 3-7

Jean-Léon Gérôme's *Pollice Verso* painting

Source: Dea Picture Library / Getty Images

lovers.[14] Its title, Latin for "with a turned thumb," highlights the work's tight focus on the moment in a staged battle in the Roman Colosseum when a victorious gladiator stands poised to deal a death blow to his fallen opponent and looks to the crowd for its verdict. A row of white-clad women is seated nearest to the action. Gérôme's decision to depict Rome's vestal virgins as appallingly caught up by bloodlust makes the scene especially enthralling.

Jean-Léon Gérôme was an already-renowned genre artist of the Romantic movement, known for painting fraught moments set in faraway places—historically and geographically. On top of his genius for composition, his renderings of contextual details were praised for their scholarly accuracy. He took pride in his meticulous research, which drew heavily on findings in the still-new fields of archaeology and anthropology.

But what really made the difference for Gérôme was that he was represented by an agent who had elevated the selling of paintings to an art in itself. Adolphe Goupil, founder of the gallery Goupil & Cie, was a highly entrepreneurial dealer who understood early on how the emerging technology of photographic reproduction would transform the market for art. The two began working together in 1859, and sealing the partnership, Gérôme married Goupil's daughter Marie in 1863. There was even more serendipity involved here in that as the head of a business with international offices in London, Berlin, and New York, Goupil had the market power to put his ideas into action. The handful of firms that, like his, could pay artists handsomely for paintings, confident that they could sell mass-produced color copies at under twenty-five francs apiece, quickly became "the unacknowledged legislators of the art world," as one historian puts it. "It was they who carried an artist's reputation into every home in the country and to all the four corners of the globe."[15] At the same time, however, Goupil was no fan of the impressionists or any other modern school. So it was a decidedly old-school painter—and a son-in-law—who benefited most from his business model.[16]

And it was the up-and-coming bourgeois class of art appreciators in the United States who would prove to be the richest market for such reproductions. Also disinclined to adopt works that strayed far from realism and familiar storytelling, these avid buyers brought the last element to an almost-magical combination of disruptive technology, globalizing commerce, and product with mass appeal— completing the recipe for a nineteenth-century blockbuster. As one journalist would report in 1877, "There are but few French artists of modern times whose works are more known, studied, and appreciated in America than are those of Gérôme."[17] And among those works, *Pollice Verso* was one of the most popular.

Gérôme's images, therefore, were easy to come by—the novelist Émile Zola would say dismissively, "He paints a picture for it to be photographed and printed and sold by the thousand."[18] But by the same token, their very ubiquity gave real bragging rights to whoever saw the *originals*—and especially to whoever owned them. And this is how *Pollice Verso*, like many of Gérôme's works, came to be purchased by an American tycoon eager to put it on display to the public—in the very city where Talmage was preaching every Sunday to standing-room-only crowds.

But Talmage was hardly the only cultural influencer who took inspiration from Gérôme's vivid Colosseum paintings. For example, the bestselling novel of 1880 was Lewis Wallace's *Ben-Hur: A Tale of the Christ*. In this story, the author made his protagonist a charioteer, not one of the gladiators, but the spectacular setting was the same. It would prove to be "the most influential Christian book of the nineteenth century."[19] (It was, of course, the novel that was adapted into the epic film *Ben-Hur*, which starred Charlton Heston, won a record eleven Oscars in 1959, and was seen by ninety-eight million people in cinemas across the United States.) But it was Talmage who took the thumb gesture a turn further than his contemporaries and their ancient sources. Looking at the thumbs turned downward in condemnation in Gérôme's painting, he surmised that the crowd must have used an opposite gesture—a *thumbs-up*—when they wanted to exercise mercy. His inference would change the course of American gestural culture.

But was his interpretation even accurate?

. . .

Talmage's audiences may have taken his words as gospel, but the painting he was referencing, Gérôme's *Pollice Verso*, had already stirred up an academic kerfuffle with its depiction of the Roman "turned thumb."

Everyone agreed that when ancient writers used terms like *pollicem vertere* or *convertere* ("to turn the thumb"), they were referring to a gesture meaning "put him to death" in response to an appeal for mercy. That is the unmistakable meaning of, for example, the words written by Rome's great satiric poet, Juvenal, around 100 BCE.

Deriding his fellow citizens, Juvenal notes that the dissolute men who used to make merry at the public arenas "now mount shows themselves, and kill to please when the mob demand it with turned thumbs."[20] Likewise, in a later Roman poet's account, we learn of a gladiator's slim hope when "the crowd threatens with a hostile thumb."[21] Later, in Italian, the expression *fare pollice verso* meant opposing or condemning. But nowhere in the extant literature is a description of just *how* the thumbs were turning.

Yet a painter wanting to portray a moment when thumbs were fatefully turned had to make a choice, and Gérôme made his. Classics scholars, irked by what they thought to be fake news, were quick to take issue. Monographs were published.[22] There was no *downward*-facing thumb indicated anywhere in the literature, they insisted—and even less was there any inkling of a thumb pointing *up*. Instead, to describe the gesture made to show mercy to a fallen gladiator, the texts used the phrase *pollice primo*, meaning "with thumbs pressed." As the leading Latin-English dictionary explained in 1880, "To close down the thumb (*premere*) was a sign of approbation; to extend it (*vertere, convertere*; *pollex infestus*) a sign of disapprobation."[23]

Gérôme defended his painting by citing a renowned archaeologist as his source—but his written reply to critics does not actually state what he considered the correct thumb direction. In retrospect, the two camps may not even have disagreed. While the turn of the thumbs in *Pollice Verso* is evidently downward, the other way to describe them is that they are extended *toward* the fallen gladiator—who, being on the floor of the arena, is below the spectators in the stands. One reasonable speculation about the origin of the "hostile thumb" gesture was that it mimicked a dagger being pushed to the losing combatant's throat. (The merciful "pressed thumb" could then also be understood as mimicking the opposite, a sheathing of the blade.) Thus, Gérôme's thinking might have been perfectly aligned with the purists.

And as for the gesture signaling that the fighter should live, an artifact discovered more than twenty years earlier should have headed

off further argument because it did feature a clear visual represen-
tation. What became known as the Cavillargues medallion, made of
terra-cotta and dating from the second century, had been unearthed
in 1845 by artist and amateur archaeologist Léon Alègre near the
city of Nîmes—home to an ancient Roman amphitheater that still
stands today. Alègre submitted his find to a more accomplished
archaeologist, Auguste Pelet, who published his interpretation of
the artifact. As Pelet describes the medallion, which was afterward
placed on public display, its bas-relief design clearly depicts the end
of a battle between two types of gladiator always pitted against one
another, a *retiary* and *secutor*:

> After a long and fierce struggle, the outcome of the fight still
> remains uncertain, but the athletes mutually gave proof of so
> much courage and skill that the people, represented by four
> figures seen in a gallery, grant them both the exemption, mis-
> sio, the greatest favor that a gladiator could obtain.
>
> This action is perfectly indicated on our artifact by the ges-
> ture that the people used to make in these circumstances: to
> express one's will, one raised a hand laying the thumb under
> the fingers when one wished to save the life of a vanquished
> gladiator. But here both athletes have deserved popular favor,
> and the cries of Pugnantes Missi (the fighters have their leave)
> join this gesture of benevolence.[24]

Talmage was wrong, in other words—there was no thumbs-up in
the Colosseum. And Gérôme may have been right, but he misinter-
preted the gesture—there was, technically, no thumbs-down, either.
This might explain why neither gesture was in use before their re-
spective blockbuster influences. In fact, an unbelievably thorough
compendium of oratorical gestures published in 1875 contains no
mention of any turn of the thumb.[25]

Yet thanks to Gérôme and Talmage—and the new media technologies that reproduced their work rampantly—the two gestures

and phrases took hold in the American vernacular so quickly and pervasively that there would be no weeding them out. Isa Blagden, a novelist of the mid-nineteenth century, once wrote, "If a lie is only printed often enough, it becomes a quasi-truth, and if such a truth is repeated often enough, it becomes an article of belief."[26] That's how it was with the story of the thumbs-up gesture.

In a way, though, it seems fitting that the American middle class got its way on this—that the matter of how a hand signal works would be determined not by the educated elite but by the boisterous crowd—the crowd effectively gave a thumbs-up to the thumbs-up gesture, historically accurate or not. Whether the arena comes courtesy of Marcus Aurelius or Mark Zuckerberg, showing the thumb is a perfect vehicle for sharing sentiments and swaying outcomes in a vast and noisy space.

In chapter 4, we move from the history and iconography of the like symbol to the biology of clicks: what happens in our bodies when we experience likes.

Chapter 4

Why Do We Like Likes?

As the head coil was placed over her face and as her body began to slide into the bore of the fMRI (functional MRI) machine, the young woman—let's call her Hannah—might have had a moment of misgiving. Maybe it hadn't been such a good idea to respond to that message board post offering compensation to be part of a research study. So far, the protocol had been fun enough. A few weeks earlier, she had been asked to go through her Instagram account and select several photos she had posted recently. And soon after checking in at the lab this morning, she had been told she would spend the session using an interface designed to mimic that social media platform to respond to images submitted by the other sixty or so participants—all of them between the ages of thirteen and twenty-one. Now, she was flat on her back, her head in a kind of grid, and under orders not to move a muscle—except for the finger she would use to press the like button.[1]

Hannah was having her brain scanned, and the point of this study, led by Lauren Sherman, was to shed light on "the neural mechanisms underlying individuals' experience with quantifiable forms of peer endorsement on social media." In other words, Sherman wanted to know if your brain lights up when you hit the like button—and if so, what parts of it. Sure enough, her research shows there is activity, and a lot of it. The findings align with patterns already observed in how positive social attention offline goes straight to the brain's "reward center."[2]

This was just what we were wondering about as we headed into this chapter. Why exactly do people like the like button? Why did the world take to it like a duck takes to water? What are the psychology and brain science behind it? What itch does it scratch?

There's no question that people do like the effects of the like button. The uptake of this simple digital tool was so immediate and astonishing that it quickly became the most used feature on social media—and by now is at the fingertips of the majority of humans on the planet.[3] Statistically speaking, you've probably already pressed it several times today. If you're under twenty, even more. Perhaps you recently posted a photo on Instagram, a short video on TikTok, or a message on Facebook and have been eagerly checking as the like count rises on your phone—each new like accompanied by a buzz in your pocket, a flutter in your heart, or a small smile as you look to see who else has just joined the ranks of your admirers. Or perhaps you liked the content of others you admire or with whom you identify or want to curry favor.

There's also no question that when the like button was introduced, no one had to be instructed in how to use it. Usually, a new technology comes with a user's manual—why does this one require no how-to guidance at all? Finding the answers took us to roots below the level of consciousness, back to the bases of early hominids' survival and the selection dynamics of evolutionary psychology. As we'll recount here, we learned that people like likes because these symbols tap into the deep human propensities that turned us into *ultrasocial beings.*

. . .

We talked to Nicholas Christakis about what is going on here. A preeminent voice in the field of evolutionary sociology, Christakis conducts research at Yale University and writes bestselling books probing the question of whether there is a fundamental, innate "human nature" that unifies us and distinguishes us from other species.[4] He believes there is, and it consists of the elements of our psychology that predispose us to engage in social activities. Together they make up a "social suite" of instincts that compel humans to seek out and benefit from interaction with other individuals. A very long evolutionary process has yielded the human trait that he calls *hypersociality.*

In a dog-eat-dog, survival-of-the-fittest natural world, why would we develop such socially oriented instincts? Christakis explained that it's partly because of the immense survival benefits of *social learning.* This is our superpower as humans, the feature that has allowed our species to attain its commanding heights. All animals learn and evolve as they interact with a challenging world, from experience gained through trial and error, but only certain primates have the advantage of the ability to learn from the experience of *other* individuals.[5] To the extent that early hominids were capable of social learning, they upped their chances of survival by avoiding the mistakes and emulating the successful moves they saw others making. Natural selection meant that these capacities were replicated and that the following generations could make even fewer ignorant blunders and even cannier choices.

But social learning doesn't come cheap, in an evolutionary physiological sense. It calls for more cognitive power to make social connections through various capacities—not least the spoken language—and the ability to retain accumulated social knowledge. This is why, across the Pleistocene epoch, humanity saw a massive change in its gray matter. Between 2.6 million and 11,700 years ago, anthropologists can now say confidently, the human brain got larger and larger, for a long time increasing slowly and then, about 800,000 years ago, at an accelerated rate. One theory for that spike was that the megaherbivores that had made up most of the early hominids' diet became far less abundant, whether from overhunting or climate change or both, and less brainy individuals were unable to master the more demanding challenge of surviving on smaller prey.[6] For whatever reasons, the average human adult brain went from the forty cubic inches in our earliest ancestors to a peak of ninety-two cubic inches some ten thousand years ago. (Interestingly, after that point, which marks the beginning of the agricultural revolution, the average brain size shrank back to eighty cubic inches. Perhaps, once the concept and methods of growing one's own food were figured out, humans were more able to survive and reproduce with less brain matter as seeders and sowers than as hunters and gatherers.)

It all sounds good, but our brain's increasing size and complexity were something of a luxury good. Once we got to the *Homo sapiens* stage, our brains were tipping the scales at unprecedented weights (a human brain today weighs on average 2.98 pounds, while the brain of our nearest genetic relative, the chimpanzee, is just 0.85 pounds). More important, in a resting state, this greedy organ was consuming fully a fifth of all the oxygen taken in by our lungs. In a landscape where it took nearly every ounce of an individual's energy just to be sufficiently nourished, warm, and healthy to reproduce, the expansion of brain size became an expensive proposition.

In the long run, however, this uniquely human variation turns out to be a real bargain, because of the efficiency it brings to the learning process. Being able to watch someone else wield a tool in a novel way and think "I see what you did there"—that saves a lot of time and effort. Knowledge acquired from others' experience is an incredible evolutionary shortcut: we no longer need to actually jump off a cliff to know that it's not a good way to try to fly. We can deduce this from the result experienced by the last person who tried it. Better still, with language, we don't even have to have directly watched that ill-fated effort. We can hear about it or perhaps wait to see a cave wall picture. As learning becomes connected and social, the survival benefits of a big brain quickly outweigh the physiological costs.

But something more than language capacity is required to advance social learning. Our extra brainpower also allows us to put a finer point on just whom we are learning the most from. This brings us to the pervasive and well-studied phenomenon of *homophily*, or our preference for people recognized to be similar to ourselves. It's a term coined by sociologists Paul Lazarsfeld and Robert K. Merton in the 1950s in a paper they wrote about two communities in Pennsylvania and New Jersey.[7] Having conducted extensive studies of these two populations, they found that everyone was fundamentally inclined toward social interaction with people resembling themselves— unfortunately also bringing about segregation. Just as birds of a feather flock together, they concluded, people have a subconscious and deep-seated predilection to gravitate and respond positively to sameness. A vast amount of research since has established how powerful an influence homophily is in many social settings.[8]

This homophily trait is tightly connected to humans' powerful advantage of social learning, because we see people like us as the best proxies for ourselves. When we observe others who are similar to us, we have higher confidence that their experiences are relevant to our own journey through life. We learn faster by watching the forays of these highly credible test pilots and seeing the outcomes that they enjoy or endure or that bring their end. Homophily, through this evolutionary lens, is a way to save ourselves a lot of effort, trial and error, and painful or fatal learning experiences. More-efficient learning increases survival odds, and therefore homophily is rewarded through the natural world's brutal selection processes.

At the same time, there is an evolutionary advantage in also having some measure of *heterophily*: the desire to be around and learn from those who are *not* like us. This is where we get complementarity that extends our abilities. It's also how novelty is introduced into our thinking and how we gain new skills, perspectives, and knowledge. It's the judicious combination of a lot of homophily with a soupçon of heterophily that allows us to get the optimal survival-enhancing learning benefits from the others around us—essentially a mixture of proxies and pioneers.

Both have social learning benefits, and people have different preferences for how much they stick to their own kind and how much they interact with people they perceive as different, but the overall effect is that homophily becomes naturally ubiquitous.[9] The easiest way to go through life is by hanging with people who are like you, because the lessons they learn are more obviously more relevant to you.

Highly related is our capacity for friendship—a special part of the "social suite" that allows us to extend our reciprocity and trust beyond our kinship circle. We might think of it as the strongest form of homophily: friends are the people with whom we have the most in common, who laugh at the same jokes, who seek out the same pleasures life has to offer. And between the learning benefits of homophily and the trust benefits of friendship, humans gain greater powers of cooperation—the inclination and coordinating capacity to work together toward shared goals that we cannot achieve as individuals working solo.

Another human strength is our affinity for what has been called a "mild hierarchy." The idea is that hierarchy is useful for a group of individuals to achieve a shared goal—it saves time and energy to have someone acting as the leader providing directions others will follow. But in the animal kingdom, these alpha group members achieve their preeminent positions by force and maintain them by threats of violence. The walrus that aspires to be head of the group must take all comers and, better yet, dissuade any pretenders to the throne with memorable shows of force. By contrast, mild hierarchy is a kind of social sorting and an arrival at productive leadership based not on who can coerce others but on who is perceived by others as the group member from which they learn most. In a mild hierarchy, members of the rank and file gravitate to individuals they see as the best sources of useful knowledge, in light of what those standout performers have done and how well they share the lessons to be taken from their experience. This is, Christakis explained, why politicians want to be photographed surrounded by people: the group shot signifies that they are perceived not as aggressors but as bestowers of value. By contrast, it can be dangerous for an individual of other species to approach a higher-ranking member since, in a coercive hierarchy, proximity means a greater likelihood of taking your knocks. Part of our human blueprint is that we want to stick close to higher-ups because of the learning benefit.

There are other corollaries to an evolved social learning capacity. They include our habits of expressing gratitude and acknowledgment—essentially saying "thank you for letting me learn from you, dear similar and more experienced person." Because we are happy to get pointers about threats and to gain opportunities, we learners receive the imparted knowledge in a fashion that will encourage more of it to come our way. And in turn we as givers respond to that flattering affirmation—it's what keeps us sharing so actively what we discover.

As most of us know, all of this give-and-take is taking shape most actively in adolescence. Decades of real-life research has clearly established that people in their teens and early twenties are inordinately concerned with liking and being liked. The challenges of forming and deepening social relationships, and the prized state of having many friends, take on heightened importance during this phase of life. For many, popularity is central to self-esteem, feelings of well-being, and a sense of identity.[10] But at no age do people grow out of the basic instincts just described. And because the power of social learning has all these psychological orientations connected with it, operating at a subconscious level, we don't have to deliberately pursue it. In fact, we are helpless to do otherwise, assuming we are normal in our psychological makeup. (And if we are not, we get diagnosed as antisocial and serve as an exception that proves the rule.)

To sum up, why do we like liking and being liked in general—offline, that is, in real life? It's pure evolutionary psychology: it boosts survival chances. And how does it boost survival chances? By contributing to the homophily, friendship, communication, and mild hierarchy that maximize social learning. We have evolved to get a buzz from liking, and that applies to both getting likes and granting them. And this is the very deep psychological territory the like button was able to tap into. To check a post and see that some likes have accumulated since it was posted causes a little frisson of delight, and to see a piece of content and decide to bestow a like on it, knowing it will matter to the person who posted it, sends a tiny feel-good surge through the nervous system as well.

. . .

This brings us back to Hannah and her neurotransmitters. The purpose of the brain scans she and her fellow research participants underwent was to find out what's actually going on to produce that buzz. What neuroscientific plumbing creates it? Researchers were positing that the pleasure of engaging in hypersocial activity was not just a cultural artifact—a value passed from civilized generation to civilized generation. That would imply that such activity would have to be, and could only be, reinforced by human-made systems and deliberate intentions. Instead, it seems, these predilections come preinstalled. The impulse to engage in hypersocial behaviors is embedded in our brain's essential wiring and operates in the subconscious realm.

As Hannah was scrolling through the content that she and her sixty peers had posted on Instagram, the research team was recording what things she chose to like and how her brain was responding to her own and others' use of the like button. One behavioral pattern became immediately apparent: the participants were more apt to like photographs that had already received a relatively high number of likes from others. Another was that "viewing photos with many (compared with few) likes was associated with greater activity in neural regions implicated in reward processing, social cognition, imitation, and attention." Most strikingly, the team saw activity across a wide variety of regions in the brain when, as Lauren Sherman reported, "the teens saw their own photos with a large number of likes."[11]

Across the eons of human development, what has been selected for in the evolutionary process are not just anatomical traits—the equivalents of the giraffe's neck or a certain finch's beak shape—or even larger brains per se but also behavioral habits and their underpinning physiology. Genes affect the structure of our bodies, but they also affect the neural processes that shape our social behaviors.

All our behavioral urges, perceived sensations, and emotions are activated by the workings of neurons in the central nervous system. Each neuron has receptors responding to neurotransmitters, hormones, and other signals that tell it whether to fire or not. This neural circuitry has evolved to support our choices to engage in certain behaviors that benefit us and to discourage harmful behavior.

Neuroscience is still a relatively young field. In one celebrated study in the early 1950s, James Olds and Peter Milner implanted electrodes into the brains of rats and rigged up a Skinner box: a setup with a lever that would, whenever a rat pressed it, directly stimulate a selected portion of its brain.[12] The objective was to discover if there were specific areas that served as reward centers and aversion centers, and if so, where exactly they were. If passing current through an electrode stuck into a certain spot caused the rats to respond by trying to repeat that experience, this repetitive behavior would be a clear indication that the sensation was pleasurable and that a reward center had been located.

The results were almost unbelievable. The rats instantly learned the effects of pressing their levers, and those whose electrodes were stuck into the septal area of the brain were obviously having the best time. They proceeded to do so up to seven thousand times per hour, to the extent that they could not be lured away from their levers by favorite foods when they were hungry or by water when they were thirsty. Male rats could not be deterred by access to a female rat in heat. Neither did they hesitate to cross over a surface they knew would shock them painfully to get back to the bar. The septal area is now known to consist of several subgroups of neurons and to contribute heavily to the emotion-driving network of structures called the limbic system. Further work by Olds then produced similar behavior in rats when the site targeted for stimulation was the nucleus accumbens—which neuroscientists have implicated in the response to both natural rewards and addictive drugs via the release of dopamine.

Dopamine is a neurotransmitter (or chemical messenger) used by the nervous system to send messages between nerve cells. More specifically, when we do something pleasurable, it acts as a feel-good neurotransmitter. What kinds of stimuli cause dopamine to be released in humans? David Linden, a professor of neuroscience at the Johns Hopkins University School of Medicine and chief editor of the *Journal of Neurophysiology*, sums it up this way: "Most experiences in our lives that we find transcendent—whether illicit vices or socially sanctioned ritual and social practices as diverse as exercise, meditative prayer, or even charitable giving—activate this pleasure circuit in the brain. Shopping, orgasm, learning, highly caloric foods, gambling, prayer, dancing till you drop, and playing on the Internet: They all evoke neural signals that converge on this same brain circuit activated in Olds and Milner's rats."[13] The same thing is going on when someone gets a kick from cocaine—or nicotine, marijuana, or alcohol. All these substances affect this dopamine-mediated pleasure circuitry.

So, as Hannah's fMRI scanner tracked her neural activity while she clutched her like button apparatus and engaged in liking various content she saw on her screen, the scanner was picking up the downstream effects of little dopamine shots—pleasant buzzes in her brain nowhere near as potent as what an artificial electrode could deliver but nonetheless in their quieter way teaching her to repeat the behavior. Such is the motivating power of the dopamine mechanism; it is the common currency for providing small, instant rewards for many behaviors and experiences that serve us well (in moderation, at least)—including those involving social connections.

Take, for example, the findings of a study in which participants were presented with various inputs while their brains were scanned using fMRI.[14] When, according to the study protocol, a member of the research team paid a participant a seemingly offhand and sincere compliment, key areas of the recipient's brain were activated in just the same way they were by the receipt of a monetary gift. There's a reason, evidently, that we call it "paying a compliment": the brain's response to this bit of social warmth matched its response to getting cold, hard cash.

But remarkably, dopamine isn't only released at the point that we get the valuable thing. It isn't just an after-the-fact feeling of gratification because something desired has been received and consumed. The real power of dopamine is that the neurotransmitters are released in *anticipation* of the reward. We get a shot of dopamine when we imagine getting the desired thing and are first forming the idea to go out and make it happen. At that point, signals from stimulated neurons cause dopamine release, activating what is known as seeking behavior. We are hit with some level of craving for the neural response we recall enjoying the last time we engaged in the same behavior.

Dopamine's role in boosting our desire for the stuff we need is very useful to human survival because in nature, this stuff is usually scarce and takes real work to attain. We have to go out of our way to do the planning and exert the effort to get it. In an extremely popular podcast, Andrew Huberman, a professor of neurobiology at Stanford University, interviews brain science experts and discusses in lay terms how their work can be applied by ordinary people trying to enhance their personal well-being. As he sums it up, "The simplest way to think about dopamine is that when our dopamine levels are elevated, we tend to focus our attention on outward goals—the things we want—and we feel motivated to pursue them."[15]

So there are really two dopamine spikes—one at the thought of the desired thing and the other at getting it—and this means there will always be some degree of "reward prediction error" involved. The joy of realizing a goal will not perfectly match the joy that was anticipated from that outcome. But this is the essence of learning through experience. Knowledge is accumulated constantly through the brain's updating of its predictions according to the mismatch between initial predictions and outcomes. If things do play out as expected, then the baseline is reinforced. If the result is surprising, by being either more gratifying or less so than predicted, then the baseline is revised.

This means that the first shot of dopamine, experienced when the baseline prediction of an outcome is envisioned, sets up the second shot to be potentially extra gratifying in the moment of attainment. It may produce the pleasant surprise of a better-than-expected result. But importantly, even if the result is disappointing, the disparity will yield a learning benefit—and the brain will adjust the baseline expectation to a level that is easier to attain the next time. Anna Lembke, the psychiatrist and addiction medicine expert behind the bestselling book *Dopamine Nation*, explains how this double shot acts on the mind of the gambler.[16] Placing a bet activates the reward-anticipation dopamine release, and then winning (when it happens) activates the reward-response dopamine release. And to the extent that the outcome is unpredictable, the dopamine release that comes with a win is even more gratifying. But Lembke also reports, "My patients with gambling addiction have told me that while playing, a part of them wants to lose. The more they lose, the stronger the urge to continue gambling, and the stronger the rush when they win— a phenomenon described as 'loss chasing.'" She quickly follows up with another point: "I suspect something similar is going on with social media apps, where the response of others is so capricious and unpredictable that the uncertainty of getting a 'like' or some equivalent is as reinforcing as the 'like' itself."

When we talked with Lembke, we asked her to elaborate on this thought. Did she think the rapid spread of the like button was because of its ability to tap into the brain's reward centers? Her answer was, in a word, yes, but her response deserves a fuller quotation:

> We are deeply social creatures. Moving around in tribes has allowed us to survive millions of years of evolution and even thrive. Being in groups allows us to steward scarce resources, protect ourselves from enemies, and find mates. So, we're driven to making connections as a matter of survival, and we make those in a variety of ways: We make them through managing and organizing ourselves in hierarchies. We make them through having sexual relations or other physical contact. We make them by having the same emotion as another human being. And we make them by expressing liking or regard. And since, from an evolutionary perspective, it's very important that we move in groups, our brains get us to do that by releasing feel-good neurotransmitters like dopamine and serotonin and oxytocin. When they're released, it feels good, and they essentially prompt or encourage us to make more connections.

. . .

So we have seen that digital liking taps into exactly the same neural reward mechanisms as pleasurable experiences in our physical, offline lives. Does that mean that the digital like is identical in every respect?

Lembke pointed out to us one crucial difference: the inclusion of the like tally, counting up and displaying how many people have registered a thumbs-up on a piece of content. "That quantification in particular," she told us, "is something that dopamine is very sensitive to and seems to augment the reinforcing and also the addictive potential of these media." Essentially, Lembke says, "what social media has done is taken human connection and *drugified* it by distilling it down to its most essential reinforcing properties."

This goes straight to the larger thesis of Lembke's book: we in the United States (along with many other countries) have become a "dopamine nation" because our civilization has reached the point that the gratifying stuff that causes our brains to release this feel-good neurotransmitter has become the opposite of scarce—much of it is plentiful and requires little effort to obtain. And because the shift from scarcity to abundance has happened so fast, our brains have yet to adapt to it. As we constantly indulge in what life has to offer, still following the instincts that served our ancestors so well, we are now overstimulating our reward circuits.

Huberman, in a 2020 episode of his podcast, spoke with holistic physician Kyle Gillett about dopamine rewards and the dangers of overstimulation.[17] Gillett's intriguing analogy was to compare the repeated spurring of dopamine responses in compressed periods to a wave pool at a recreational water park. In a wave pool, the waves are of different sizes, and if there is a higher peak, it is followed by a deeper trough. Similarly, when your dopamine release is exceptionally high, you lose "almost all the water from the wave pool, and then, when you crash from that, not only is the trough low, you have less dopamine in the pool than to begin with." It's an analogy he uses often because dopamine is very sensitive to exciting stimuli, and therefore "the depth of the pool can change very quickly." Drugs are the most injurious form of this overstimulation because they goose the brain to release dopamine even in the absence of any of the stuff we've evolved to need and want. And they therefore derail the pursuit of those things, much as Olds and Milner's lab rats starved themselves as they pressed their pleasure bars.

. . .

We started this chapter with a simple question: Why do we like likes? We wanted to know why liking took off in such an accelerated way. What made it so appealing to people, and what made it so intuitive? Why did nobody ever have to be instructed in how to use this innovation and how to get value out of it? On one level, the answer is as simple as can be: we like to like because it gives us a little frisson of anticipation and satisfaction. It gives us that shot of dopamine. This is true of the IRL form of liking, and it is just as true of hitting the like button on a digital interface.

A little less simple was the background to that buzz: the ultrasocial psychology that is being maintained by our brain chemistry. We are biologically predisposed to make social connections that lead to the sharing of information. That feeling you get when you're granting or getting a like is the button tapping into many thousands of years of human development. As Nicholas Christakis told us, "The like button is built, in a very deep and distant way, on the back of evolutionary biology."

This is the key, then, to the meteoric rise of the like button in its applications and daily use. There are other staples of user interface (UI) design across software—like the back button, scrolling, the ability to swipe left or right, a click to open a new window, a field for search—and it's hard to quantify the relative usage of interface features like these. But surely the like button ranks at least near the top of the most used interface features in the world. It's just not possible for something to become so popular *without* connecting with something fundamental to our biology. As described earlier in the book, Yelp developers were initially surprised at how effectively a one-click positive feedback mechanism could encourage people to spend their time and energy submitting opinions that they felt could benefit others. Now that we understand that humans have evolved the neuro-anatomical and neurochemical machinery to support social learning, it doesn't seem so surprising. As a superefficient way of acknowledging and rewarding social information shared, the like button went straight into our evolutionary veins.

What the answer boils down to is the human superpower of social learning, spurred on by brain-level mechanisms to encourage friendship connections, cooperative action, mild hierarchy formation, and communication. We like likes because they give us a rush as our brains release dopamine, and we feel this rush when we give and get likes. On a deep level, the brain perceives the like as a small pointer to a lesson to be learned from someone else. It recognizes that the people most worth learning from are those whom other people appear to be learning from—that there is a mild hierarchy being established—and that the learning from people who are *like me* are probably the most relevant to me. The like serves as a micropayment for this appreciation—a token of recognition and gratitude—saying thank you for letting me socially learn from you, dear similar person, or much-learned person.

Each feature of the like that appeals to you now is a part of liking that has always appealed to you. It's no wonder that all those parts of the brain light up when social media users are subjected to fMRI studies. Really, we should not be surprised that the like button taps into all of this. The very old machinery of the human brain is perfectly adapted to respond to the very new machinery of platforms like Facebook, TikTok, and Instagram.

But how exactly is that machinery gathering and moving these tiny bits of social information along? This brings us to the topic of our next chapter: the like button as a triumph of coding and what happens to digital likes inside the man-made silicon brain of social media.

Chapter 5

What Happens When You Click?

One morning last week as you tucked in to your breakfast, you also checked your Instagram account. Scrolling with your left hand as you ate with your right, you made your way through the stream of photos and short videos uploaded lately by people you follow. At the top of the feed was a friend's new puppy, comically thwarting an attempt to train him. So cute. You instantly clicked the heart-shaped icon. Further down, it was a guy you knew in school, marking a milestone in his newfound fitness journey. If that didn't deserve kudos, what would? And next came one of those "unboxing" videos where an ecstatic customer gets an order delivered—in this case, a colorful, oversize stainless steel water tumbler. You hesitated for a moment because of course it was promotional content . . . but you *have* been on a hydration kick lately, and the woman's commentary *did* resonate, and that cup's design really *was* on point . . . Oh, well, why not—you gave it a like, too, before moving on to ponder a meme posted by Snoop Dogg.

And then, earlier today, such an interesting coincidence: you went to a family gathering and approached your sister just as she was taking a huge swig from a plastic water bottle. "Gotta stay hydrated," she said and laughed as she wiped her mouth on the back of her wrist. "But these plastic bottles—what a waste. You should see the great stainless steel cup I just ordered." And she went on to name the very brand you'd noticed last week.

Now it's occurring to you: maybe that wasn't a coincidence. It could be that you inadvertently spurred her purchase. You and your sister follow each other on Instagram, and you occasionally see her name at the top of a post that tells you the content was "liked by" someone. Presumably, your name sometimes pops up on her screen, too. When you tapped that button last week, did it cause the same video to show up in her feed, and did your sort-of endorsement encourage her to go and buy a tumbler?

This chain of events is the other side of how the like button has impact. In the last chapter, our curiosity took us to inward territory; we were trying to understand the psychological mechanisms of liking. But here we consider how the likes people dispense go out into digital machinery and cause things to happen in the world. You might think of the click of a like as a moment of data creation that simultaneously gratifies the human brain and also shapes the digital domain outside the individual. Thanks to computer storage and analytics, individual likes are stored, aggregated, mined, and deployed for all kinds of purposes. And while we as authors were certainly aware that internet activity is constantly tracked and that the user data compiled by social media platforms is the key to their value and everything they do, we wanted to know a lot more about how it all works and how those capabilities are evolving. How does this tiny act of kindness travel from the *front end* of the process—a minimal-effort tap at the UI level—to the *back-end* machinations of data centers and communication networks so vast they now demand a meaningful percentage of global electricity use? This chapter will clarify the journey of your click inside the machine, in a way that is hopefully understandable to non-coders.

The path of inquiry starts with a simple, clever program—a half page of code authored one day in 2005—and quickly takes us into the realm of social media algorithms combining like data with many other inputs to accomplish myriad tasks. Most obviously, platforms like Facebook, TikTok, and Twitter use these algorithms to provide personalized news feeds and timelines to their account holders, and to sell highly targeted advertising opportunities to marketers who want to see their products and services appear in those settings. The platforms also use the algorithm in search results, profile pages, and conversational spaces being frequented by desirable customers.

The algorithms are the powerful drivers of what gets seen and what goes viral in a social media network. But they need vast amounts of data to work with, and they get most of that from their own users' actions. The algorithms' procedures for prompting more sharing, reposting, and liking—and their equations for crunching the resulting data into smart decisions—are highly proprietary. As we'll see, one big issue is that much of this methodology is a black box, and this chapter will ask whether it must remain one. But we're getting ahead of ourselves. For the moment, let's focus on the transparent, first answer to what happens when you click—the performance of the straightforward tasks you expect to be done when you hit the button.

· · ·

On the topic of the like button's front-end functionality, we talked recently with Russ Simmons, the Yelp cofounder and CTO. As discussed in chapter 1, his creation of the code for that site's *useful*, *funny*, and *cool* buttons was a major milestone in the development of the like button. When Russ wrote those original lines of code, it was the first time a web user could visibly add an appreciative reaction to a piece of user-generated content (in this case a review) simply by clicking once and without enduring a page refresh. Yelp cofounder and CEO Jeremy Stoppelman remembered the day well, so we asked if he might look in the Yelp code repository to dig up that old, simple piece of code for us—and he did (figure 5-1).

On the day that Russ sat down to write this code, he was aware of Gmail's use of JavaScript to create a very sophisticated front-end with no page reloads. However, such an application of JavaScript had yet to be widely understood and adopted at the time. So from his point of view, it was cutting-edge and exciting to apply this technique to the problem at hand. He said he wasn't certain it could be done, because it had never been done before, but he'd give it a try. For his colleagues, it became more obvious what a tricky technical puzzle this was when they saw Russ put on his headphones to code and even take the beloved office puppy, Darwin, off his lap.

Darwin, by the way, despite his noninvolvement in this task, would soon enough make his own enduring contribution to the business. He was calm and sweet-natured, as Hungarian Vizslas tend to be, but he *was* a puppy. Just a few weeks later, the Yelp site

FIGURE 5-1

Yelp code for adding a reaction to a piece of user-generated content

```
// Submit new feedback
this.set_button_image(which, 'press'); // change to the new image
set_element_html('review_feedback_message.' + this.rid,
    '<span style="color:gray;">Saving...</span>'); // show a message
this.state = STATE_SUBMITTING; // record that we are submitting (so we don't double-submit)
this.setting[which] = true; // record the feedback locally
this.stats[which]++; // increment our local count for this feedback type
set_element_html('review_feedback_stats.' + this.rid +
    '.' + which, '(' + this.stats[which] + ')'); // display the new feedback count
// Send a request to the server to record the feedback
// append the nocache argument to prevent Safari (and other browsers?) from caching request
callURLAsyncNotify('/review_feedback?rid=' + this.rid + '&amp;fb=' +
    which + '&amp;state=on&amp;nocache=' + (new Date()).getTime(), state_change_cb(this));
```

Source: Courtesy of Jeremy Stoppelman / Yelp

inexplicably went down for Russ, sending him scrambling to find the fatal glitch. But he quickly realized that it was just that his internet wasn't working, although everyone else's in the office was working just fine. At that point, Russ looked over to see Darwin sitting on the floor, happily munching on the ethernet cable that ran to his laptop. After that, Yelp redesigned its standard "404 error" page to include a photo of Darwin, looking quite sheepish. Curiously, a year later, Amazon started putting dogs on its error pages, and plenty of other sites have since followed suit. Was this just an organic trend as puppy after puppy caused an outage? The folks at Yelp never did the research to find out, preferring to celebrate Darwin as a trendsetter. Such is the evolutionary nature of site development on the web.

Back to the code, once Jeremy Stoppelman unearthed the early version of code that we had requested, we asked Russ Simmons to walk us through it line by line. How exactly did that set of instructions take a small piece of positive feedback from a cursor on the screen through a user's laptop (there were no smartphones at the time) to Yelp's servers?

"OK," he started out, dragging his index finger along the first line, "so when you click the mouse button, a USB cable carries a packet of information to your computer, and its operating system receives this. It knows that your web browser is the foreground application, and so it passes this information—that there was a mouse click—to the web browser. The web browser receives this click and then tries to figure out which element of the page the mouse cursor is currently over. In this case, it knows that it's over the *useful* button, and so the web browser sends an event to the client-side code saying that this page element received a mouse click."

By "client-side code," Simmons was referring to programs installed and running on the user's machine—programs such as the web browser itself—as opposed to those running in the cloud, which is referred to as server-side code. So this is "JavaScript code that runs in the web browser that is authored by Yelp in this case and has in advance told the browser, 'Please notify me if any of these elements are clicked on.'" In technical terms, the code has "registered an event listener" so that it will be informed of click events on relevant elements. "When the browser is told by the operating system that a click happened," Simmons continued, "it looks at what elements the mouse cursor was over, sees that one of those elements was one that the JavaScript code had registered a listener for, and it notifies the code, 'Hey, you were interested in clicks on these elements. A click has happened.' The code then basically makes a request to the server to register this vote, if you will, and that request includes authentication information." In other words, the server registers the clicker's unique user ID, along with the review that was clicked on and which of the voting buttons, *useful*, *funny*, or *cool*, had been chosen.

What about that unique user ID—how does that happen? Or at least, how did it work in 2004? "It's not exactly like this now," Simmons noted, "but in the old days, the server gave you a cookie, which is a persistent piece of information in your browser. That cookie has some secret tokens in it that prove to the server on subsequent requests what user you are. Every time you do something on the site, that information gets passed up to the server, and it proves that you are the same user that was given that secret earlier when you gave us your login and password." Now that the server trusts that you're, say, Bob, it lets you write a review or leave a feedback vote as Bob. This ID check is happening on the server side the moment that piece of information comes in, and once it's been validated, the event is logged by the server in what is basically a table in the database capturing in its different columns all the related information: the type of vote or feedback it was, the ID of who clicked the element, the ID of the review it was responding to, and when this happened. (By now, it is an unfathomably large table.)

Simmons then moved on to the part of the code that switched the appearance on the user's screen of an element after it was clicked. "That's pretty straightforward," he explained. "After sending that request to the server, the JavaScript code would then just switch the button image to a new image, with the new image's green color denoting that the button was in the clicked state," at the same time "changing the shading to look like it went from convex to concave."

This color shift meant the code needed another wrinkle, as well. Simmons explained: "Whenever a user loads a page—say, a business page or a user page—for every review that appears on that page, Yelp's server has to check that table to see if this user has already voted on any of them and show those votes in line." Clearly, it's a desirable feature for a user looking at a list of reviews to be able to see what feedback they have already given and know that it has been stored. But providing for this capability, Simmons said, "is actually a lot more work than registering the votes. If you load a page with a hundred reviews, you have to do this query that many times, checking in this table for past votes on any of those hundred reviews."

This tallying feature adds up to a fair number of instructions but not too many lines of code. Simmons recalled that day two decades earlier, when he had taken an afternoon to create it: "I'm going to guess it was in the range of twenty to thirty lines."

So that's the front-end story of a pathway that is really very simple—but that gives rise to a much more complex system because it hardly stands alone. The less obvious and more interesting things happen because, as each click is noted by the system and recorded, the action adds to aggregated data that can be analyzed for patterns. Your motivation to tap a thumbs-up or a heart icon may begin and end with your desire to show support to a content poster. But combined with the likes of many others, your simple action has power that extends way beyond your momentary impulse to reach out and touch someone. What seems like an ephemeral action produces a data point with a life of its own—a life that may last forever, working its way into endless other corners of influence and action.

. . .

The like button has come a long way since this 2005 deployment by Yelp. As it radiated through the realm of social media in the following years, it began helping sites decide who else and how many others should get a look at a given post. It helped put many pieces of user-generated content over the threshold that allowed them to get seen widely—and attract even more likes. To explain how, we need to move beyond the front-end interface and the question of how data is initially prompted and captured, and we must begin exploring what happens in the back-end realm of data storage, analytics, and triggering actions.

The first aggregate your like adds to is the platform's understanding of *you*. Whether or not the like shows up on your profile and stays there, the system notes not just the emotion you clicked but also when and in what context the response was registered. And it retains a memory of your click, adding to the pile of likes it can sift through to know you better. It may see that you often like what one certain person posts and deduce that this person is a meaningful focal point for you. Or that you like just about any post with a sloth in it. Maybe you just find sloths adorable? Whatever the reasons for your predilections, those patterns are filed away.

Likes are hardly the only thing adding to the platform's understanding of you. There's other information you voluntarily divulged in your profile, like your relationship status and whatever biographical details you have included in your self-description. In light of this level of user data alone, a platform can make surprisingly good guesses about what your habits and interests might be. But now add to that all the richer insights into your personality and activities that you are actively divulging without really thinking of them as self-revealing. The content of your posts is a gold mine of details, often yielding an accurate picture of, for example, your political views, your sexual orientation, and your hobbies. You also expose much about yourself in the affinity groups you join and the accounts you follow.

And there are still deeper levels of insight you are revealing even less consciously, like your physical location. In the aggregate, geolocation data gathered directly from your devices might show you to be a frequent traveler, a suburbanite who commutes to the city and works certain regular hours in a certain building, or someone with a family you join for the holidays in Kansas. Even more revealing might be the photos you upload, making it clear, perhaps, that you are a dog owner or a theatergoer, as well as whatever your selfies might reveal about your ethnicity, body mass, and fashion preferences. The language you post in, the vocabulary you use, how bad a speller you are—all add to the platform's knowledge of you.

Revealed behavior, to use the term market researchers favor, has always been the best predictor of future choices. It's the real deal: no matter what you might answer to a question like how frequently you buy toothpaste or whether your tastes in film run more to action hero or art house, the truth can be found in what you actually do. Your use of a platform can reveal behavior you don't even reflect on or recognize about yourself. The timing of your activity on the site, for example, can say a lot about the typical rhythm of your day, and the emotional valence of your postings can serve as a kind of mood ring you didn't know you were wearing. Like poker players highly adept at spotting tells, the social media platforms you use pick up on signals you're not conscious of transmitting. The public got a glimpse of this near omniscience in 2017 when a Facebook marketing team in Australia spelled out more than its management would have liked about how deeply it understood the continent's young people. According to the team's report, passed to a reporter for the *Australian,* Facebook's monitoring of users' emotional states could enable delivery of ads at "moments when young people need a confidence boost," for example, or when they were feeling "worthless," "insecure," "defeated," "anxious," "silly," "useless," "stupid," "overwhelmed," "stressed," or "a failure."[1]

This is the kind of capability that a social media user's every tap of the like button helps construct as back-end algorithms perform functions based on patterns in the data. To be sure, liking is not the only way users betray their feelings—and it might not even be a particularly honest mechanism. Does it count as a revealed behavior that you like all your boss's LinkedIn cerebrations? Probably the number of seconds you linger on the page or whether you post a comment reveals more of the truth. But at the least, your likes have a lot to say about your priorities, the relationships you value, the identity you are working to project.

Social psychologists refer to the "avowed norms" an individual knowingly expresses as opposed to the norms an individual has adopted less consciously or deliberately. Sometimes the norm avowed—like "we must always put the shopping cart in the corral in the parking lot"—matches the norm observed, and sometimes less so. But even in the breach, the avowal is an indication of what you believe is good. In this sense, likes can be seen as little restatements of avowed norms. A social media platform may know that you are still checking your ex's feed rather obsessively or lingering long over pictures of juicy cheeseburgers, but it also counts as valuable information that you are liking tips for finding new love and how-to videos for vegan living. In fact, the tension between your avowed likes and your revealed behavior may be the richest vein of insight to mine about you. And unless you have a long-standing relationship with a very good therapist, your social media platforms may know more about that than anyone.

. . .

If you use a social media platform with any frequency, its understanding of you doesn't stop with you—it goes on to comprehend your social attachments. From the standpoint of social scientists, this is what made the platforms of the Web 2.0 era such a bonanza for their field. "As individuals bring their social relations online," one group of researchers noted, "the focal point of the internet is evolving from being a network of documents to being a network of people, and previously invisible social structures are being captured at tremendous scale and with unprecedented detail."[2] To a network theorist, you're just another kind of node.

And to a social network analyst, what's most important is how you fit into a system of connections or relationships. People or algorithms trained in such analysis can apply graph theory to map the social structures by which individual nodes connect to each other and interact through links (known in the trade as edges) or influence each other through adjacencies at their edges. The nature of the connections mapped by a social graph can vary. A *follow graph*, for example, reflects connections made between fellow nodes as individuals. Using this graph, someone could find the path from you to any other user of the system through everyone's choices of which accounts to follow. (And it might well be a short two-node trip through footballer Cristiano Ronaldo, with his Facebook followers topping 170 million.) But in a *like graph*, a node-to-node connection is based not on links but is formed when two users reacted in the same way to a piece of content. If both you and some stranger halfway around the world liked the same nature video, for example, the graph shows a direct link between you two, even though no connection was intentionally made by either of you.

Undoubtedly you have heard about degrees of separation, if only through the familiar claim that no human on earth is more than six friendship links from any other. Mapping degrees of separation is a fundamental element of social graphing. If you're a social network analyst looking at a graph, separation is the number of *edges* you can count in the shortest path between two nodes.[3] And another fundamental task in social graphing is the calculation of *clustering coefficients*, indicating how intensely nodes are connected. In other words, having found a group of birds of a feather, the clustering coefficient indicates just how tightly or loosely they flock together. So every time you press the like button, that action also assists the platform's mapping of your like graph as it works to gain a more comprehensive view of your tastes, your influences, and whom you influence.

Chapter 4 discussed the natural human tendency toward homophily, or people's inclination to spend more time with those they regard as similar to themselves. That preference of nodes to connect

to nodes of the same type is what social media platforms have been
built to serve and to accelerate and intensify. When Mark Zuckerberg
stood up at a Facebook developer conference in 2008 and touted the
marvelous potential of Facebook's social graph, most people even in
that tech-savvy audience had probably never heard the term before.
But he was right about the power of enabling homophily at such unprecedented scale. "The social graph is changing the way the world
works," he said. "We are at a time in history when more information
is available and people are more connected than they ever have been
before, and the social graph is at the center of that."[4]

At that point, according to the company's founding engineer
Jeffrey Rothschild, the company had about eight hundred servers
containing about forty terabytes of user data.[5] Seven years later,
Facebook was storing more than three hundred petabytes of data
in its Hive, and processing six hundred terabytes of incoming data
every day.[6] Since then, the stores of data have only grown, along with
the incentives for social media companies to gratify the preference
for homophily rather than its opposite, heterophily—the desire to
encounter and interact with those who seem different.

. . .

Why does so much theory and energy go into understanding you as
an individual and as a social creature? All this data allows the code
to *curate your digital experience*. By helping a platform construct
your like graph, your use of the like button influences what shows up
most prominently in features making recommendations to you, such
as whom to follow and, most importantly, your feed.

A little prehistory on the feed. Facebook gave us this *feed* ter-
minology in the social media sphere when, on September 5, 2006,
it launched its news feed feature and, along with it, the mini-feed
that individual users saw on their pages. This was a substantial shift
in direction, given that up till then, Facebook had been all about

profiles. In the profiles era, your page showcased your own posts and you used it to maintain your social network. If you wanted to see anyone else's posts, you had to deliberately choose to visit their profile—their content would not be pushed to you unbidden. But outside social media, the term *news feed* goes back further, borrowed from the news industry, which, since the mid-nineteenth century, featured news agencies like Agence France-Presse, the Associated Press, and Reuters. These were suppliers of news gathered through outposts around the world and conveyed to client newspaper publishers and broadcasters. The news feed was the continuous stream of printed content that spilled out of a clack-clacking teleprinter in a subscribing newsroom. On that end of the wire service, editors would curate their own selections to serve their local audiences.

The feed pioneered by Facebook was news not in the sense of articles published under journalist bylines about current events but rather the news of your world—the updates posted by people you chose to connect with and your own posts. But in keeping with the news model, Facebook put the emphasis on what was newest. And because the feed was chronological, posts got buried pretty quickly by newer content. Facebook and other platforms soon realized that this approach wasn't serving users well. It would be better to base the content on a more meaningful prioritization, ideally with an understanding of what a given user would not want to miss.

This approach brings us to the realm of algorithms. The word *algorithm* is part of nearly everyone's vocabulary, but the idea is still intimidating for many. It shouldn't be daunting: an algorithm is just a sequence of steps laying out the detailed process or instruction set for reaching an intended outcome. The recipes you consult to prepare your dinner, the steps you take to change a tire, and the instructions you study as you assemble a piece of DIY furniture are all (nondigital) algorithms. In this case, the recipe is written out in computer code and the steps are being taken by a software program—and the outcome is finding the most delicious content for an individual social media user. In software, an algorithm is a set of commands that must be followed for a computer to perform calculations or other problem-solving operations—a finite set of instructions carried out in a specific order to perform a particular task. The word *algorithm* stems from a ninth-century Persian polymath, known as Al-Khwarizmi.

Many elements go into understanding who you are. Likewise, all these factors influence what you might find intriguing enough to click through and view. The question is, What weight should each element be assigned relative to the others? What's the optimal recipe? What's the objective?

The primary goal here is to *keep you engaged*. That's why all this energy goes into understanding you and what you like. It's to maximize your engagement, to keep you glued to the screen and interacting with what you find there. Everything else that social media does to produce value flows from this effort to engage. Having you maximally engaged is what constantly feeds the beast. Engagement generates the data the whole apparatus and business model depends on. So job one is to encourage you to keep interacting, displaying emotions, and expressing preferences both overtly and unwittingly. This goal, by the way, is why most social media companies decided not to provide for any one-click disliking of content. A dislike button might seem like a natural thing to include, and at first, it was not uncommon for a site to have a thumbs-down or downvote button. But people are so turned off by negative feedback that its usual effect is to decrease engagement levels overall. On that basis, most sites that had the feature soon dropped it.

If you're highly engaged, then you are doing more than tapping the like button. You're posting, reposting, commenting, DMing, following, unfollowing, and more. But as a baseline indicator of engagement, the like button is hard to beat. By repeatedly performing this low-friction, prosocial, feel-good action, you keep pinging the platform, indicating that you are still present, still have a pulse, and are still paying attention to content. And indeed, your sustained use of this minimally taxing function habituates you to interacting on at least some level with the platform. And that makes it more likely that you will step up to richer forms of interaction.

At a fundamental level, feedback has long been collected from media audiences and used to curate content. For more than half a century, television broadcasters paid for Nielsen data, initially from paper diaries and later from a succession of electronic devices, to help them answer the question "Is this working with the audience?" Using the signals they received, broadcasters adjusted their programming choices and airing schedules to grow their viewership. But thanks to like button clicks and other data, social media sites can now make such adjustments at a much more granular level. The question is no longer how the audience in general is responding— it's how the individual is responding. And this much more precise targeting and adjustment happens much faster and more frequently because the feedback loops spin at the speed of electrons, not at the speed of paper surveys being filled out, mailed in, and analyzed by humans.

Also very different from traditional feedback mechanisms is the rich, interactive nature of social media—the fact that users not only respond with observable actions but can give voice to their feelings. For consumers to talk back to an advertiser in the twentieth century, they needed to write letters saying, for example, "Your food is terrible and the portions are too small." And their interactions with each other on a product's merits were literally word of mouth, the oldest form of advertising. But in the social media era, that two-way communication is enriched, amplified, and made much more seamless and rapid, and all of that is empowered by the ability to understand those signals. Even if you had ten million raw likes, it would be quite hard for a human analyst to see the patterns. So the algorithmic innovation to understand this granular data is a big part of the story.

The algorithm a platform devises affects what rises to the top of your feed. And because you will never be able to scroll through all of it, its design also means that you are not being exposed to certain ideas, or themes, or people. When you check your Instagram, you may perceive it as just a straightforward feed, but in fact you are experiencing a highly curated digital experience.

. . .

Not surprisingly, given how important a platform's algorithms are to the quality of experience it provides to users—and the quality of ad targeting it provides to marketers—every platform owner keeps its algorithms close to the vest. Fernando P. Santos is a researcher who has devoted untold hours to ferreting out the details behind some of them. He spent some time with us describing what he knows—beginning with the fact that the various disclosures made by platforms about "how our algorithm works" barely scratch the surface of that question. When we asked where we could find flow diagrams showing how various elements factored into social media algorithms, he quickly disabused us of that hope, explaining that Facebook, for example, had published only system cards and some blog posts and that even a recent June 2023 disclosure provided "just basically an intimation of what are these algorithms predicting and what are the key inputs."

Model cards are a recent innovation, a standard that many in the AI community prefer for documenting the models they build and use in terms understandable to people who aren't machine learning experts—like the policy makers and regulators working to make AI models more fair, transparent, and explainable. A card is usually a one- or two-page plain-English description of a model's scope and performance. Santos gave credit to X (formerly Twitter) under Elon Musk's ownership for going further than other platforms to describe its workings, but even for this company, Santos said, "the code itself is known, the different components are known, but then, how they are related—that's not obvious."[7] Dashing our hopes to divulge all in this book, Santos summed up the situation: "I would not say that there's a platform that I know that is fully transparent."

At least we do know what a social media algorithm looked like in the earliest days, just after Facebook introduced its news feed. As we described, the feature had been launched with the information simply presented in reverse chronological order: new stuff was at the top, and older stuff followed (and quickly got buried). But in light of feedback from users, Facebook quickly realized that this system wasn't optimal. In 2007, Facebook came up with a more sophisticated algorithmic approach for feed prioritization, which it called EdgeRank. Developed by Serkan Piantino, it essentially built on the basic approach behind Google's PageRank algorithm to prioritize content according to three objective measures: affinity, weight, and decay rate. *Affinity* is how much a candidate piece of content had been engaged with by other similar users (including its like count). *Weight* adjusts for the type of content (video, comments, or other format) and how well the type matched the user's viewing habits. The *decay rate* refers to how quickly a piece of content is becoming uninteresting or irrelevant to other users, as indicated by their liking and sharing of it. (Typically, a piece of content sees three-quarters of the total impressions and reach it will achieve within two to three hours of its publication.) Putting these three elements into an algorithm with relative weights assigned to them allowed Facebook to automate the production of a feed based on relevancy to the individual user rather than general popularity or, worst of all, simple chronology.[8]

It worked pretty well, although Facebook stopped using the algorithm in 2011, when it started to use machine learning algorithms.[9] But its early success didn't mean it was the right algorithm for another social network or even another part of the same platform. There are multiple algorithms at work on, for example, Instagram, depending on whether you are looking at your feed or at reels content. The design of the algorithm depends on the objective.

Here's what we know about the weightings at TikTok. In 2021, the *New York Times* acquired a leaked document called "TikTok Algo 101," prepared for internal use at that company by its engineering team in Beijing. "The document explains frankly," the newspaper reports, "that in the pursuit of the company's 'ultimate goal' of adding daily active users, it has chosen to optimize for two closely related metrics in the stream of videos it serves: 'retention'—that is, whether a user comes back—and 'time spent.' The app wants to keep you there as long as possible."[10]

At Instagram, according to its CEO Adam Mosseri, each feature bases its own rankings on the use of that feature. The feed algorithm looks at your activity in-feed (likes, comments, shares, and saves) and information about the post and the person who posted. The stories algorithm focuses on stories you've viewed or engaged with in the past to serve up similar ones. The explore-page algorithm seizes on your activity in explore (likes, comments, and shares) and information about the post and the person who posted. And the reels algorithm factors in activity in reels (likes, comments, shares, and saves) and information about the reel and the person who posted.[11]

Another example is YouTube. Initially its algorithm prioritized videos according to sheer clicks, indicating views selected. Later it switched to favor videos that rack up longer viewing time (the time people spend watching before moving on to something else).

You're probably noticing that a big part of the algorithm isn't your own input. It's also factoring in what the people you affiliate with, admire, and resemble have liked. You've made clear whom you consider to be your tribe, so it is a reasonable assumption that you will respond in similar ways to things they like. Their likes affect your feed.

But even beyond the people you see as like-minded, there are people you don't know who, at a profile level, look like you. So stuff that they like also makes its way into your feed. This is classic marketing segmentation but taken to the very precise and personalized level. It's casting you not as a member of a cohort numbering in the millions, like suburban women, but as a member of a cohort of perhaps a thousand or whatever minimal number might be worthwhile for a marketer to target in light of the size of the selling opportunity. At whatever scope the circle is drawn around people who look like you, their likes affect your feed.

By the same token, your likes affect their feeds. Your likes and theirs go into the giant aggregate. And the algorithms become more complex and sophisticated. Facebook's EdgeRank algorithm has long since gone by the wayside. Today's newsfeed algorithm factors in many variables and draws on your whole history of interacting with the platform.

It's still true that any like you register will have its most potent influence on your own feed—on yourself. But from there it will affect the feeds of your most engaged friends, then your wider network, then the people you don't know but whose interests overlap with yours. The poet Rilke wrote that "I live my life in widening circles that reach out across the world."[12] In the same way, we express our likes, and in widening circles, they affect others in lands we may never see.

· · ·

Meanwhile, for the person whose content you reacted to, your like is aggregated with likes by others . . . indicating to them what kinds of content are most popular. And it factors into their engagement score along with other engagement-based metrics such as comments, shares, and clicks. If the sum is higher than other pieces of content, then that post sends a high-engagement signal to the algorithm, indicating that the post is popular. That signal allows it to gain more visibility. This is why a post can blow up—the design of the system ensures that reach will escalate rapidly. If the sum is relatively low, the opposite happens, of course, and the post sees limited reach.

By the way, research shows that more-positive posts get more sharing (the negative stuff gets *faster* sharing but ultimately not as much).[13] And Facebook's internal research on site usage revealed that when the like button was hidden, users interacted less with posts and ads.[14] So, likable stuff that is visibly liked earns the greatest engagement.

We'll see in chapter 6 how much the measuring of engagement matters to those in the new influencer industry because it determines what influencers get paid for a shout-out. But it isn't just people who are making a go of being a paid influencer who get scored in this way. The traditional influencer gorillas, the consumer goods advertisers like the Procter & Gambles of the world, are also tracking these metrics. We talked to Dan Gardner, one of the ad industry's pioneers in this realm, about this. He told us that the like button allowed people "to reprioritize content based on a ranking of what is relevant. And that was a huge moment of change from a business and advertiser perspective, because then you had to think about the *effectiveness*, not just the *reach*, of your *following*, but the *effectiveness* of your *individual pieces of content* to make sure they rank properly."

There is now a whole industry that has grown up around assessing and comparing the engagement levels achieved by different brands. "In the default Unmetric formula, comments and replies are weighed higher than favorites and likes because comments and replies start a conversation. Shares and retweets are weighed higher than comments and replies because shares and retweets intentionally amplify audience reach of a tweet."

And at the other end of the spectrum, if you are just an average social media user who posts the occasional piece of content, a platform has assigned *you* an engagement score, too. Whether your next post will appear to the people in your network, let alone anyone beyond it, is highly dependent on that score. And your engagement score is highly dynamic, always reflecting how your content has performed lately.

. . .

In the scenario at the beginning of the chapter, we depicted a situation in which you wondered, "Did the like I casually clicked on a social media post influence my sister's decision to part with some of her hard-earned money for something she could have lived without? If so, how did that happen? Where did that like go now, and then later, to exert its bit of added influence?"

To get to the answers, this chapter began with a detailed look at the surface of an application: how the interface of the app captures the like and knows what to do with it. As we showed, the like button—already arguably the most used feature of social media interfaces and still spreading—is a social tool for users but is also fundamentally a piece of code. We explained how the pieces of data are created when you tap a like button and how they are stored.

We then probed what is known and not known about how such data is analyzed and what actions are informed by it. We described how all those atomized moments of emotional reaction wend their way through the coded universe of social media. We explained what users are divulging, from an information systems view, when they tap the like button. We showed how this information feeds into a like graph charting connections between an individual and other people who like the same things and into a social graph created by a platform to define you more fully as a node in its network. We looked at the history of news feeds and how liking affects the content you and your friends see. We traced the many lines of impact that every like has; the lines go far beyond the momentary affirmation experienced by the specific people whose content was liked.

To be sure, the like also does just what you intended and has its greatest impact closest to home. It makes a positive difference to the person you give it to as a direct affirmation and message of encouragement. It gives someone you care about that little rush of dopamine. It adds to their self-confidence and makes them that little bit happier. But as we revealed, when you use the like button, you are sharing quite a bit about yourself, perhaps more than you imagined. The little, multifaceted communication of information is added to all the other pieces of personal data you have sent, creating aggregate answers to questions such as these: What interests do you have? What does your lifestyle contain? Are you conservative or liberal?

This information affects which ads and influencer marketing pro-grams you'll be targeted with and the order of content in your feeds. Your like is also combined with billions of likes by other people to allow patterns to be discerned at higher levels. All of this helps many people—and machines—make decisions large and small. It's a giant octopus of an apparatus today, to which every like contributes its minute influence.

By now, thinking back to your sister in the scenario, it must have occurred to you that the direction of influence might have been the other way around and that her interest and research into tumblers might have caused you to be shown the post. Would the conversation today with your sister have happened if you hadn't liked that post? What else has occurred in real life among you and your friends be-cause you tapped that button? Is there any way of knowing?

We imagine the process as a light rain beginning to fall on the surface of a still pond. Each droplet causes ripples, and the concen-tric circles spreading across the water increasingly overlap. It's the same for all the times you've clicked a like button. The tiny waves of influence begin at the interface of the app, flow to the code base processing them, and collide and overlap in vast lakes of data stor-age. Through a mysterious and complex process that no one could retrace, all those taps of the like button combine in ways large and small to change the course of the future.

In chapter 6, we'll examine how this vast lake translates into eco-nomic value as amassing, analyzing, and monetizing likes becomes big business.

Chapter 6

The Business of Likes

R yan Detert used to live for likes, but now likes are providing him with a better living than ever. Back in 2013, he had amassed a following on Twitter of over thirty million people checking out his content on travel, fashion, style, and automotive news—and was figuring out how to capitalize on that audience by making deals with agencies and brands to promote what they were selling. Now he runs the world's largest influencer marketing company, as founder and CEO of Influential, an agency that matches up consumer brands with the right people in today's vast army of social media tastemakers, most in their twenties and thirties, doing what he used to do.

Detert remembered clearly how he came to start the business he runs now: "This all started off as me, as a creator of content, having a very difficult time monetizing my audience, and realizing that in order for me to scale and get dollars from agencies and brands, I had to provide more." Across his various exchanges with marketers, he had learned to appreciate their needs for assurance around media measurement, for example, and brand safety. He wished there was a trusted data provider that would allow marketers to see when it would make sense for them to work with him instead of some other influencer. That's when he saw the opportunity that shifted his energies toward a new objective: "I decided to build that technology, to get them that checklist that they need, as opposed to being the greater influencer myself."

Detert is now a key player in a whole tier of service providers in what some people have called the like economy—the new commercial world of people and brands on social media amassing thumbs-ups and finding ways to be paid for that positive attention. Talking with him and others in it, we couldn't help but wonder, Just how big a deal is the like button in terms of economic impact? How did this little UI feature give rise to a whole new industry while also transforming old industries full of big, long-established firms? How did it fuel the rise of the social media platforms, mobilize billions in capital, and drive the growth of some of the largest companies in the world?

As we discovered, likes have turned into revenues for players of all types and scales and in every corner of a fast-expanding social media economy. *Fast Company* puts it this way: "In terms of sheer impact, the like button was one of the most successful pieces of code ever shipped."[1] And note that this is a business magazine weighing in. *Fast Company* isn't talking about personal psychology, political influence, or social dynamics here. It's talking about money.

. . .

One especially interesting aspect of the tale of the like economy is that, when the like button was invented, no one realized how lucrative the data it creates would be. In those early days, when it was still just an article of faith that social media sites would somehow generate revenues someday, user base size and average time on platform were everything. And for some sites like Yelp, the immediate concern was how to motivate user-generated content. Adding a like button really moved the needle on all these fronts while conveniently providing a metric to make all this user data known and comparable across sites.

A standard observation today is that it has since become a standard observation that data has become the most valuable commodity in the world—it's the "new oil"—and the like button creates vast amounts of it. It's a standout example of what Stan Davis and Bill Davidson, authors who envisioned the emergence of the information economy, called "information exhaust." They assert that, as a by-product of operations, many businesses generate information that they don't bother to retain or do anything with, but that this information often can and should be seen as the raw material of new products and services. Writing in 1990, their favorite example was the magazine *TV Guide*, which back in 1953 had seen that there was real money to be made just by compiling programming information from all the television networks serving a given locale and presenting it in an integrated, easy-to-consult format.[2] A corollary is that a business seeing an opportunity to create an offering based on information exhaust might offer customers a product or service for free just to get the data from their use of it. When the like button was invented, no one was thinking along these lines, but the liking data quickly added up to just such raw material.

To start with, like data is exceptionally clean, valid, and analyzable. For example, when the researchers behind a study focused on how well computers could assess people's personalities needed to figure out the right data source to test their hypotheses, they landed on Facebook likes. "Likes represent one of the most generic kinds of digital footprint," they determined. "For instance, liking a brand or a product offers a proxy for consumer preferences and purchasing behavior; music-related likes reveal music taste; and liked websites allow for approximating web browsing behavior. Consequently, Like-based models offer a good proxy of what could be achieved based on a wide range of other digital footprints such as web browsing logs, web search queries, or purchase records."[3]

Like data is also superabundant. Pressing a like button is such a lightweight act that, by 2015, Facebook reported its users were doing so on roughly four million posts every minute. This was more than enough to feed the machine learning technologies of new analytical fields.

Likes also have the advantage of being intentionally registered feedback that shows up in real time. And the data can even be location based, thanks to the mobile devices people use to register their likes as they go through their daily lives. All in all, like data is full of insights to be mined. It powerfully predicts whom we'll vote for, what levels of education we have attained, and what we will choose to buy.[4] However unwittingly, a small tool designed to promote user engagement and content creation has turned into a font of invaluable data.

. . .

The enormous amount of data gained by tracking account holders' activity is what drove the early social media platforms to their watershed moment—the short period in the mid-2000s, when nearly all of them shifted to ad sales as the central feature of their business models (figure 6-1). Today, the advertising revenues coming into Facebook, X, Instagram, and the rest add up to hundreds of billions of dollars.

These investments are justified by the rates of click-throughs and view-throughs they generate for marketers. Click-throughs are the number of people who click on a sponsored post, which takes them to a website to see more, and view-throughs are the number of people who, having been exposed to the post, go on to visit the sponsor's site later. Knowing that someone has liked similar posts in the past is valuable because it raises the likelihood of their taking those actions. This person is a good target for messages that in a social media environment can be precision-delivered.

FIGURE 6-1

Evolution of global ad spending over the last 40 years, by medium

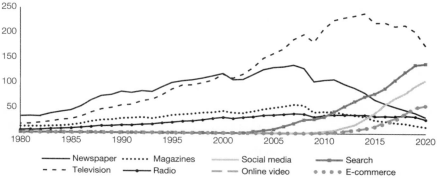

Global spending in billion USD

Source: Adapted from Raconteur, "Ad Evolution" infographic, October 2020, https://www.raconteur.net/infographics/ad-evolution/.

Years ago, David Edelman, now of Harvard Business School, published with his colleague Richard Winger an influential article called "Segment-of-One Marketing."[5] The thesis was that the information environment was fast evolving to make more data available about individual customers, and if a marketer had the right information, it could tailor a unique value proposition to any given person. Martin heard David present his ideas in 1989 and remembered that they seemed interesting but unrealistic given the state of online business at the time. But this was an information problem that would soon become solvable, as interests became more discoverable and the costs of communicating with customers fell. Recently we asked him, Is that essentially what has happened?

His answer was unequivocal: "That *is* what happened." And how did the like button play into that? "Audiences," he said. "The like button aggregates audiences, who are defined by their affinity for what they are liking. So as a marketer you either use the poster—the creator who actually is being liked—as a *channel*, as an influencer, or you pay to place ads next to their content to target the people who like them. Those people have a profile of someone you want to target because they have expressed that they like someone else. So it's audience aggregation."

Nowadays, we might easily forget that there are other social media business models besides selling audiences to advertisers. But things could have gone in another direction. To this day, in fact, LinkedIn doesn't bring in as much revenue from ad sales as it does by charging the job seekers, headhunters, and employers who are its most avid customers for extra bells and whistles. Instagram, too, held out for at least a few years before it introduced advertising options, even though by the time it launched in 2010, the ad-based business model already dominated the social media scene all around it.

Helping to nudge the platforms toward ad-sales models was the fact that online advertising was already well established outside of social media. As early as October 1994—just a few years after the advent of the web—*Wired* magazine had run the world's first banner ad on its site, promoting AT&T. And according to Joe McCambley, one of the people involved in creating the ad, this very first foray was an astonishing success: "Of those who saw the ad, 44% clicked. Not only did people love the experience, they loved it enough to share it with friends. We were blown away. . . . We were on to something."[6]

When Facebook launched in February 2004, Mark Zuckerberg already had some inkling that ad sales would be needed to keep his fledgling business afloat. At a point when its user count had just leaped to seventy-five hundred, he was quoted by Harvard's student newspaper as saying, "It might be nice in the future to get some ads going to offset the cost of the servers."[7] It didn't take him long to act on that thought, since the very same year brought Facebook's first project to drum up revenue from marketers with an interest in his college-student user base. It was called Flyers, because the idea was to post local ads that would be seen by account holders from specific schools, much as, in real life, they would see flyers posted on campus bulletin boards and utility poles.

A year later, in 2005, Zuckerberg started thinking bigger, hoping to tap into the much larger budgets of big businesses. He did this by offering not a standard ad buy but a more integrated kind of opportunity, by which a brand could exclusively sponsor a Facebook group, which Facebook would in turn promote to users. The deal was that the marketer—Apple was the first to bite—would pay one dollar for every user who joined its group. Apple named its group Apple Students, and anyone joining it could register for weekly raffles to win a free iPod Shuffle, get product discounts, and download iTunes songs. It turned out to be a nice deal for Facebook, as Apple Students attracted a hundred thousand members in just a few months.

Then the site introduced Facebook Ads in November 2007, which it described as "an ad system for businesses to connect with users and target advertising to the exact audiences they want."[8] By now, Zuckerberg was pursuing the real potential of promoting on social media. What Facebook offered was, he proclaimed, "a completely new way of advertising online. For the last hundred years, media has been pushed out to people, but now marketers are going to be part of the conversation. And they're going to do this by using the social graph in the same way our users do." The program analyzed data on interests, age, and location drawn from not only Facebook but also some forty other sites whose user activity could be tracked by means of partnerships and shared features.

So Facebook was offering targeted advertising buys to marketers well before it introduced its like button in 2009. But the addition of that new feature made a big difference. As Harvard Business School's Leslie John recently advised corporate marketers, "One reason Facebook advertising can be effective is that a brand's social media page reaches a highly desirable audience; likes illuminate a path for targeting ads."[9] Facebook brightened that path even more by encouraging other websites to adopt its like button feature as a plug-in, with the result that, as users liked things elsewhere, their interests would be shared back to their Facebook profiles. This interface gave Facebook even more data to analyze people's sentiments and preferences across many realms of activity and to provide more precisely targeted advertising opportunities.

The effect of the like button was to goose the company's ad-sales potential in a way that investors sat up and noticed. Consider the change in Facebook's value as a company before and after the introduction of the feature. In 2008, although it wasn't a public company yet, analysts watching venture capital deals estimated Facebook's total valuation to be $15 billion. The next year, the like button was introduced, and a year after that, major investments by Kleiner Perkins and Goldman Sachs implied a total valuation north of $50 billion. By 2011, Facebook was earning $1 billion in profit on $3.7 billion in revenue, most of it from advertising. In 2012, the year of its initial public offering, the company's market cap would top $100 billion.

Not every adventure in advertising paid off for Facebook as it experimented with different options and launched at least one expensive flub. Under that program, it told marketers it could sell them *sponsored stories* driven by its account holders' use of the like button: whenever someone liked that company or one of its offerings, Facebook would spread the word, alerting all their Facebook friends that the person had done so. The logic was simple—the program would digitize the greatest form of advertising in the world: word of mouth. But the scheme got the company sued by users who successfully argued that it violated the Canadian Privacy Act forbidding the use of anyone's name and photo in an ad without their explicit consent. The company should probably also have figured out that their users would balk at the practice, because word-of-mouth advertising, as a personal act of information sharing, tends to be highly targeted. The fact that someone likes a pricey gadget might be something they're happy to tell a college roommate but might not align with the image the person is trying to convey to the thrifty aunt pitching in on their tuition. Ah well, social media platforms live and learn. Facebook went on next to launch an *advertising exchange* by which a marketer could place a bid to get placements in the feeds of users who had previously liked something relevant to its product offerings. The idea was along the same lines as Google's ad auctions based on people's web search histories.

There's a beautiful thing about operating models that center mostly on automated functions like matching bids to ad placements. They can scale very far, very fast. Soon, Facebook realized it could serve the small-business tier with targeted advertising. Those sellers could find new customers by creating Facebook pages and buying ads to appear on well-targeted users' pages. Called Facebook Fit, this selling initiative ran workshops for owners of mom-and-pop businesses, teaching them how to take advantage of the platform's treasure trove of data. (The 2013 movie *The Internship* starring Vince Vaughn and Owen Wilson centered on two resourceful misfits' work on a similar program at Google.)

The advent of the iPhone and other smartphone devices brought another huge boost to promoting goods on social media because it suddenly created the world of mobile marketing. By 2012, Facebook was developing the capability to do location-based advertising, allowing brands to base sales pitches on real-time awareness of where users were when they posted updates or otherwise interacted with the platform.

Facebook's journey from no ads to simple banner ads to a cornucopia of highly targeted advertising options is a striking one, but hardly the only one. Twitter and YouTube also introduced ad options early in their histories and added features of their own. YouTube created a partnership program recognizing the importance of its video uploaders by giving them a share of the revenue from ads placed alongside their content. Twitter came up with promoted tweets, which invited marketers to create tweets just as the site's users did (and under the same message-length constraint) and then have them show up at the tops of their followers' feeds and even in the feeds of selected nonfollowers. Promoted tweets were the tiny social media equivalents of the advertorials that magazines had long allowed companies to place in their pages for a healthy price. They exposed the reader to a message that, because it blended in with surrounding content, was more likely to be viewed. Twitter's marketing scheme allowed the company to take its ad revenue to $583 million in 2013.

The advertising options have continued to multiply across the whole social media ecosystem. Facebook launched carousel ads and a program it calls custom audiences. LinkedIn introduced Linked In ads, and Instagram made video ads and story ads available. By 2021, Twitter had added Twitter Amplify, follower ads, and take-over ads to its promoted tweets and was reporting that 89 percent of its $4.5 billion income that year came from advertising services—a 40.5 percent gain over just the previous year.

Clearly, the switch to advertising-centered business models has paid off for the platforms, and much of the momentum was facili-tated by likes. As they made their pitches to brands, platforms had a huge advantage over traditional media by being able to present hard evidence to back up the value proposition of their targeted adver-tising opportunities. In June 2012, Facebook wrapped up the "quiet period" after its May 18, 2012, initial public offering with a splashy announcement of research conducted by Comscore showing impres-sive sales increases for the typical marketer running both paid ads and unpaid marketing campaigns on the platform. The *Wall Street Journal* covered the announcement, along with quotes from Ford and Coca-Cola marketing management crowing about the results of their Facebook ads. The report was actually the second in a multi-part series issued by Comscore and Facebook. And the title of that series? "The Power of the Like."[10]

In sum, Facebook and the other early platforms may have started out almost like toys, uneconomic hobbies, but they soon became big businesses. How? By the transformation of unpaying users into the product. More precisely, through their activity, users create data that permits them to be targeted by advertisers, creating a monetizable service for social media platforms. Media scholar Brooke Erin Duffy told us that she likes to teach her students the concept of the audience commodity, first described in the 1970s by political economist Dal-las Smythe. Like any transaction in a commodity market, audience attention can be sold to marketers in the quantity they wish to buy, at a price that fluctuates with market conditions. The commodity in question has no say in who buys it, nor does it see any of the proceeds from the sale. This last part was galling to Smythe. "He essentially argued that when we're watching commercials, when we're watching TV, we're actually doing work of consuming the advertising content," Duffy explained. "And so, even though we think of that as leisure activity, that's actually a form of unpaid consumer labor."

The same can be argued about tapping away at the like button. Every time we use it, we create value that ultimately, in aggregate, will accrue to the platform, not us. Meanwhile, as it encourages more unpaid content creation by others, it also creates the inventory of available ads. An increase in publishing by a greater number of motivated content creators means more posts against which ads can be run. Even more profitably, every time we register a like, it serves as a meaningful signal of expressed preference—a little show of warmth that a heat-seeking brand can zero in on.

It's a neat trick on the part of the platforms—giving their product away free to users without being too explicit about the fact that, in fact, those users *are* the product. Unbeknownst to most of them, they are being packaged up into blocks based on their likelihood to engage and sold to one marketer after another. Yet even those who are highly aware of this arrangement seem mainly unbothered by it. Arguably, the experience of having goods and services pitched to you is more palatable when those ads have been more precisely targeted and a greater proportion of them feel relevant. According to one survey published in 2024, around 78 percent of internet users say they use some form of social media when looking for more information on brands—and increasingly they turn to it for not only product research but also inspiration and ideas on how they could enhance their lives with the stuff being promoted.[11]

. . .

The large-scale transition we just described in social media business models was mirrored by a transformational shift in the world of marketing and advertising. Marketing scientist Christine Moorman has, since 2008, fielded an annual survey to collect data and opinions from chief marketing officers and CEOs on topics of greatest concern to them. Perhaps the biggest story between then and now has been the meteoric growth of activity and expenditure on social media and its use across an expanding set of objectives. By 2023, spending on social media campaigns had grown to represent 16 percent of total marketing budgets—and respondents indicated this would rise over the following year to 18.9 percent and, by 2028, to 24.3 percent. Companies serving consumers rather than business customers (B2C versus B2B sellers, that is) were most likely to report that these social media investments were making substantial contributions to sales performance.[12]

To understand the view of marketers making such investments, we talked to Jeff Dodds, a veteran consumer goods executive who has seen the traditional world of relatively untargeted advertising transformed by social media. He started his career as a marketer at Honda; then led efforts at Callaway, the big US golf company, and at Virgin, working for Richard Branson—all places where he was encouraged in the philosophy "that any piece of work you do nowadays has to be suitably disruptive and suitably jarring in order to get any level of traction, because there's so much marketing noise."

The problem in the early 2000s, as he remembered it, was that marketers were working with very limited abilities to see what was really getting traction. There have always been two basic dimensions to the performance of a campaign—its reach and its reception—but he and his colleagues had data on only the reach. "We were always talking about how many people have been reached by a message or a piece of communication," he told us. "But on the second axis, how many liked what they were being reached by . . ." His voice trailed off.

This second axis is what they suddenly could focus on when consumers gained a nearly frictionless way to register emotional responses: "The invention of the like button and similar technologies very quickly allowed us to get some sort of sense of whether there was a positive attribution or a negative attribution toward what we were putting in front of people," Dodds said. It was a landmark event in advertising, he said, when "we were able to not just assess how many were seeing it, but how they felt about it."

Dodds said that the very first time he saw a way to get value from likes occurred when he was working as a marketing executive at Virgin Media. "We would create ad campaigns and then put them up on YouTube and very quickly assess from the likes, dislikes, and comments on YouTube if there were things that needed to be tweaked before we produced the broadcast version of that product." It was an unbelievable source of "real-time information and very fast turnaround feedback that we'd never had before." The instant, free data he marveled at some fifteen years ago would soon bring about the utter transformation of the *attention aggregation industry*, also known as the ad business.

Anyone who has seen *Mad Men* gets what ad agencies do—at least the way things used to be. Today's agencies still work closely with marketers of products and services to develop compelling messages, and then, equipped with this creative work, they figure out how to most economically get those messages in front of legions of susceptible customers. So, the offerings of an ad agency fall into two buckets: creative work and media buying.

Until quite recently, however, both types of offerings have been more of an art form than a scientific process. As testified by every *Mad Men* scene featuring a Don Draper pitch, the creatives in an ad agency ginned up (in Draper's case, sometimes literally) alternative slogans and art treatments they thought would resonate with consumers—and the first filter they ran them through was the top management of the client company. Expensively produced ads were then placed in media that felt like the right context, given the broad segment to which the marketers wanted to appeal. In a similarly broad-brush approach, grocers and department stores compiled weekly newspaper inserts and carpeted whole towns with them. It was hard to tell whether any of this advertising was really working.

Marketers had to choose the agencies that seemed to know what they were doing and had to trust them to generate the sales.

Back in the late nineteenth century, the department store tycoon John Wanamaker famously echoed the first Lord Leverhulme's observation that "half the money I spend on advertising is wasted, and the trouble is I don't know which half."[13] It's a line that resonated with every marketer for the next thirteen decades. The only real exceptions to this problem were direct-mail promotions, couponing, and promotional contests and sweepstakes, all of which required interested customers to actively respond. With these, there was actual feedback, and marketers could gauge, albeit with significant delay and cost, which messages and which distribution lists were working better than others.

Today, digital marketing provides the same accountability that the direct-mail and other response-required campaigns of the past did, enabling marketers to tell which specific messages led to which specific sales. But the learning-and-adjusting process is exponentially accelerated. The fact that realized sales can be attributed to particular digital investments may not completely solve Wanamaker's dilemma, but it gets a lot closer to the ideal of getting bang for every marketing buck.

And now that the demand for this level of precision has become the norm, advertising agencies are hiring more data scientists than creatives. The scientists are needed to devise propensity models and write the machine learning algorithms to deliver messages customized for users' likes and other aspects of their profiles. They constantly run what are called A/B tests (that test two groups with the same offers except for one difference) by changing different variables, and they tweak the machinery to optimize the returns on ad spending. Their mission is to get the right people to take in the right messages at the right time and place for maximum impact.

To be sure, not every consumer is active on social media, and social media advertising is still not the dominant line in most marketers' budgets. But things are trending in that direction, as digital ad spending rises more generally (figure 6-2). According to analysts

FIGURE 6-2

Rise of social media ad spending

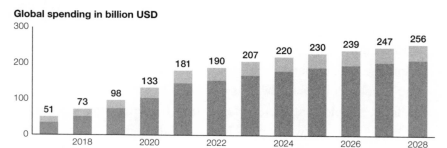

Global spending in billion USD

Legend: Social media advertising (desktop) | Social media advertising (mobile)

X-axis: 2018, 2020, 2022, 2024, 2026, 2028

Values: 51, 73, 98, 133, 181, 190, 207, 220, 230, 239, 247, 256

Source: "Social Media Advertising Spending Worldwide from 2017 to 2028, by Device," Statista Market Insights (2024), https://www.statista.com/outlook/dmo/digital-advertising/social-media-advertising/worldwide#ad-spending, BCG analysis

at Singapore-based social media consultancy Kepios, of the total amount spent globally on advertising in 2023, digital advertising consumed an amazing 70 percent. And within that digital category, the amount spent on social media is growing apace.[14]

. . .

It took a while for the *Mad Men*-style shops to get the memo. Because of their failure to adapt quickly, there was suddenly a flowering of pure-play digital agencies. Joe McCambley, who helped create that first digital ad for AT&T in 1994, did so while he was working at a new digital-focused marketing agency named Modem Media. "For a few wonderful years, while big agencies slept with their backs to the Web, we did incredible work for major brands," he recalled. "By 1998, though, spending on Internet advertising had grown to the point where the established agencies woke up. Innovative shops like Modem Media, Razorfish, and Agency.com were snapped up."[15]

It's easy to imagine that the first thing that brought those established firms into social media advertising—apart from the alarming decisions of their clients to hire other agencies—was the like button's power to signal positive sentiment. With this little thumbs-up UI feature, they gained a powerful tool. The latest research we are aware of is a 2024 paper investigating the link between Facebook likes and purchasing behavior. Taking a big-data approach and looking across eight years of Facebook use and subsequent purchasing behavior, the researchers found a positive association between registering Facebook likes for brands and then buying them.[16]

Similarly, an older study of Hollywood film marketing found a strong correlation between prerelease social media liking of trailers and other promotional fare and first-week theater receipts: "Our empirical results indicate that the prerelease 'likes' exert a significantly positive impact on box office performance," the researchers report. "More specifically, a 1% increase in the number of 'likes' in the one week prior to release is associated with an increase of the opening week box office by about 0.2%."

Findings like these—most by the advertisers themselves and not by academics—caused everything about the way that brands work with agencies to change. Valid data reflecting sentiments, usable for targeting specific potential buyers, completely upended what they were asking for and how much they would take on faith. In the days of *Mad Men*, an account manager could say "Trust us; we are creative geniuses." After the social media advertising revolution, clients needed higher proof.

. . .

We've been focusing so far on how the like button fueled a transformation in the advertising business and in the focus and operations of large firms' marketing departments. But liking is an engine of feedback that helps companies in many ways beyond targeting ads. Floating new product and service ideas, projecting social responsibility, appealing to the talent market, improving the experience of getting postsale service—all this can be enhanced according to what people indicate they like.

The rise of the customer experience industry is a prime example because it was built on an important premise: not only must businesses and other organizations attract a clientele, but they must also serve it well. Service providers focused on this listen-to-the-customer function can now do extensive analytics on data generated by people pressing the like button or otherwise expressing their sentiments as they digitally interact with a company. When marketing-focused analysts at Frost & Sullivan fielded a survey of decision-makers in the retail consumer goods industry in 2022, it found that almost 90 percent were implementing social media analytics or planning to invest in such capabilities in the coming year.[17]

This ability to gain feedback on customer service may be the even greater impact of a digital revolution that has pushed so much interaction between organizations and people into online settings. It would be the fulfillment of a prediction by Eric Schmidt, who asserted back in 1998 that "customer service is the killer app of the Web."[18] At the time, he was a tech executive at Sun Microsystems, some years before he was recruited to serve as Google's CEO. His observation may have been somewhat self-serving; Sun was just then marketing an early customer interaction system. But the vision behind that system proved to be valid and way ahead of its time. Schmidt understood that the web would radically change how businesses connect to customers. The ability to track an identifiable person's assorted touchpoints with a company across time, from website visits to retail purchases to customer service interactions—and even in some cases through their use of a connected product—has made it possible to map so-called customer journeys. With these data maps, companies can find the patterns in the customer's typical courses and experiment with interventions that might, along the way, support higher rates of customer satisfaction and retention. In other words, people pressing like buttons are helping businesses do more than just engage them more precisely as sales prospects. They are taking what has been a relationship of relatively information-poor transactions and turning it into a richer, extended conversation.

We talked to Sam Hall, who worked as a communications executive at a series of large firms—Vodafone, Oracle, and ServiceNow—over the period when social media was disrupting how corporate communications was done. "What did it mean?" he said. "It meant that suddenly people could talk back to businesses, and the advent of the like button showed a preference. And then comments opened up, and then sharing opened up, and it became almost as though there was a new currency in play and all new things could happen. The dynamics of the industry changed."

When we spoke with Tom Ward, the chief e-commerce officer at Walmart U.S., he reflected, "The widespread use of the like button made a huge impact in removing the friction of sharing at least a signal. The binary nature of it means that you have to then go and understand through more rich data or qualitative feedback or further analysis what it is that's making people like or dislike the experience. But there's some kind of beauty in its simplicity." To illustrate this, he added that during Black Friday 2023 alone, products on the Walmart website were liked by customers four million times.

Today, it's obvious that customer feedback improves operations: we all see how Uber and Lyft, for example, use passenger ratings to increase efficiency, safety, and reliability. But some of us can also recall a time—not that long ago—when the vehicle for offering such input was a wooden suggestion box attached to the wall at a company's place of business.

We also spoke with Kevin Warren, former chief marketing officer of UPS, about how that very data-oriented company sees the value of the like button. He said it's mainly about improving the quality of service in digital environments. Customer satisfaction feedback used to be infrequent and slow to arrive. "It would be surveys, maybe done quarterly or monthly, or whatever," he said. "And it wasn't specific. . . . [Now, with] real-time feedback from the customer to the vendor, we are able to test and learn different things and then adjust and test again."

For example, are the shipping options presented on UPS's website easy enough to understand and choose from? Warren shared, "We can 'test and learn' simplicity on pricing and see, if we make it simpler, whether that causes an uptick or not. Does the customer have a good experience? We can get that feedback real-time and be able to react in a more agile way." Ultimately, the use of the like button adds to the company's ability to provide truly personalized service. "Customers want their vendors or their partners to know them," Warren said. "When you can interact real-time in a digital format, there's a level of intimacy that drives additional loyalty—and that goes to revenue as well."

Every marketer we talked to made this same essential point about the value of the like button. Dodds, for example, put it this way: "What it enabled us to do was to test concepts and test messaging *at scale*—and very quickly. And even now, after twenty years, there's not really a better way of doing that than putting out a piece of information through social media to a very large group of people and asking them to give you a quick read on it, with a thumbs-up or a thumbs-down."

. . .

The prospect of putting oneself out there in a social arena and inviting a massive crowd to weigh in with a simple thumb gesture might seem rather unappealing. For some of us, the fate of the unloved Roman gladiator comes to mind. Yet it clearly does not deter America's youth. In response to a 2019 survey, 54 percent of US residents aged thirteen to thirty-eight said that if given the chance, they would become an online influencer.[19] Figure 6-3 shows the clear rise in the global influencer market.

For content creators who want to be influencers, likes are their stock in trade. They get hired by brands to promote products because of their likes history, and they perform their valuable service by liking

FIGURE 6-3

Influencer marketing on the rise

Global market size in billion USD

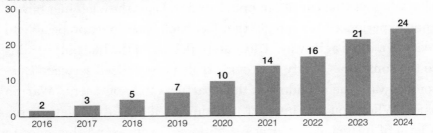

Source: "The State of Influencer Marketing 2024: Benchmark Report," Influencer Marketing Hub, September 2024, https://influencermarketinghub.com/influencer-marketing-benchmark-report/, BCG analysis

the brands' merchandise and communications. Some make astronomical amounts of money for doing so. Back in 2012, for example, a thirteen-year-old Jimmy Donaldson was an avid Minecraft and Black Ops gamer and opened an account on YouTube to post play-along videos. Fast-forward nine years, and he had become the platform's top creator, earning $54 million in YouTube income in 2021 alone. *The Street* reports, "Between $3 to $5 million each month [came] via a combination of ad revenue and paid sponsorships."[20] When we last checked, in early 2024, his average number of likes on a video (based on his fifty most recent videos) was 9.5 million.[21] You are more likely to know him by his screen name, MrBeast.

But below the stratosphere of influencers, thousands of people are making decent money off their credibility with smaller audiences of followers. Collectively, in fact, they are the biggest revenue drivers in the trade. And Moorman's CMO survey suggests that the marketing dollars will be there to pay for their services.[22] It's a topic she only began including in recent surveys, but respondents must be feeling fairly positive about their investments in the area: in the aggregate, they reported that 5 percent of their 2023 marketing budgets already involved the use of influencers in some way and they expected this percentage to more than double, to 12.2 percent, in the coming five years. Perhaps not surprisingly, the two industries reporting the highest use of influencers were communications/media and consumer packaged goods. And purely online companies are more than twice as likely to use influencers as their brick-and-mortar counterparts.

This brings us back to the previously mentioned executive Ryan Detert. His company, designed to match up marketers that want to do influencer marketing with content creators who are a good fit for their brands, uses machine learning software to keep track of literally millions of frequent social media posters, constantly reassessing who they are as individuals, how many people they are engaging, to what degree they are doing so, and what topics they are covering. It's a massive, dynamic trove of data on accounts, audiences, and content to be mined on any brand's behalf. Detert's firm, in short, serves a valuable intermediary role by connecting brands to appropriate influencers.

Other intermediaries in this new business ecosystem are more focused on serving the influencers themselves as customers. Neil Waller described to us the business he and a partner have built, an agency called the Whalar Group, which serves the community of content creators and brands that want to promote products and services through those creators. The business has multiple facets, but perhaps the most interesting is its work supporting a few hundred full-time online influencers—a group whose annual average revenues range from $200,000 to nearly $2 million.

Waller said that across the seven years since the business was launched, it had seen dramatic evolution not only in brands' interest in working through influencers but also in the professionalization of the influencers. At the beginning, he said, few of them saw what they were doing as anything more than a side hustle—perhaps 15 percent were doing online content creation full-time, and those few were usually focused on just one platform. Today, among the thousands of influencers Whalar interacts with regularly, more than 95 percent see no need for a "day job" to pay the bills. Nearly half of them have agents putting them in the way of opportunities and guiding them with advice to boost their earning potential. And they all seem to know that they can't tie their fortunes to a single platform but need to be posting in multiple places in a variety of formats. The industry has reached the point, Waller said, that he sees many content creators hiring at least one other person to help them run their mini publishing businesses, making influencing increasingly a source of new job creation.

One thing that particularly interested us was Waller's explanation of how the fees are set. In terms of what his business charges the brands it works with, the fees reflect four factors that vary with the influencer and the creative brief: cost of content production, usage rights, exclusivity, and distribution. The distribution element is itself a highly variable factor because it is driven by three considerations: the size of the audience, its level of engagement, and who the audience is. The latter two considerations can make a big difference to the revenue a content creator or influencer sees, since even a small audience can be very valuable to reach if it's in the right niche category. "Someone who is a fintech creator with a fintech audience," Waller said, giving an example, "might be able to make more money than someone who is a fitness creator." He added that, like traditional media buying, "the more targeted, the higher the cost."

How did this all get started? It began with content creators, a development that Brooke Erin Duffy, as a critic of modern marketing techniques, has tracked since its beginnings. She told us how she had sat up and noticed a 2007 online marketing initiative by the snack maker Frito-Lay when it was planning a commercial for that year's Super Bowl to promote its Doritos product. Marketers there decided to run a contest they called Crash the Super Bowl, inviting ordinary people to submit ads of their own creation, with the prize being that the top entry and possibly others would be aired during the big game. Duffy remembered that the campaign was being talked about as empowering, since in the end, some person far from the commanding heights of Madison Avenue would have their creative work seen by millions of viewers at the most high-profile media event of the year. She also remembered being outraged at that clever reframing of what she saw as a form of exploitation: "I was like, this is ingenious in terms of marketing, because you no longer have to pay your marketing staff, you no longer have to pay writers and directors. Instead, you just use us exploited dupes to do free marketing for you!"

At the same time, it couldn't be claimed that anyone was being tricked into doing that free work. If the entrants were being exploited, it was at least being done "somewhat openly." So, when a similar user-generated ad contest was launched a few years later by Unilever's skin-care brand Dove, Duffy was intrigued enough to make a study of it. What motivated ordinary people to participate in such a time-consuming, often costly competition with such low odds of winning? Was this adding to the amateur producers' power in some way, or were they, as she suspected, simply being taken advantage of? She decided to conduct in-depth interviews with a group of participants and find out.

What she learned in the process surprised her. She realized that the "binary framework of empowerment and exploitation" she was bringing to her inquiry was "completely problematic—because the reason that a lot of people were participating in these ad contests was not that they were exploited and unknowing, nor that they were completely reconfiguring power, but that, actually, they saw these commercials as a way to build *their own* personal brand." And their participations didn't even require winning, because they could display their clever entries to their own audiences. As one participant, an amateur producer named Kelly, told Duffy, "I want to be a director, and this is the third project I've done. . . . I'm hoping to use this for my demo reel, to send it out for film festivals."[23]

The first true version of monetization for any content creator was the driving of ads either through Google's AdSense or on YouTube. An enrolled creator earned money to the extent that viewers clicked on such ads and eyeballs moved directly from their content to that third-party location.

That changed dramatically with the launch of Vine, the original short-form video platform launched well before TikTok. Vine did not support clickable ads but instead called for brand integration into the content itself. This requirement presented a creative challenge: How do you in just six seconds integrate a message from Pepsi or Apple or some other brand? According to Detert, "Vine really was the genesis of influencer marketing. It was acquired by Twitter and then shelved. But not before it revealed to a first generation of influencers what was possible and gave them a big start."

Influencer marketing is quite different from the traditional use of celebrity endorsements. Yes, the accounts with the greatest numbers of followers tend to be pop culture stars from entertainment and sports. On Instagram, for example, the top accounts are Cristiano Ronaldo (595 million in late 2023), followed by Lionel Messi (477 million), Selena Gomez (435 million), and Kylie Jenner (397 million). And it's possible to hire a celebrity at this level to give a shout-out to your brand. For example, when Gomez entered into a partnership deal with the fashion company Coach and posted a photo of herself sporting one of its products and including the handle @Coach, her followers responded with no fewer than 4.5 million likes.

But the more vibrant and growing realm of influencers are people who have for the most part grown up as personalities within the social media environment. They have succeeded as audience aggregators by building their own little publishing empires, and they are attractive to marketers and other message-pushers who want to reach an audience with more authentic voices. A survey commissioned by a marketing agency in 2020 suggests those marketers are right. In the survey, a thousand American consumers were asked for their opinions about influencers, including whether they were most likely to be swayed by celebrity influencers or by influencers of another kind: aspirational, relatable, expert, "just for fun." Overwhelmingly, they expressed a preference for influencers they perceived as "relatable" (rated most highly by 60 to 70 percent, depending on product category), followed by "expert" personalities (at 50 to 58 percent). Only 17 to 22 percent professed to care most about what celebrity influencers had to say about products and services they might buy.[24]

In some influencer arrangements, the influencer does not resemble a celebrity endorsement at all, because the liking is so genuine or mutually beneficial that no compensation is required. Consider the marketing strategy of Formula E, where Jeff Dodds is now CEO. A fast-growing startup organization, its name is a reference to the renowned racing organization Formula One, because it also runs motorsport championship events, but the *E* signifies that its high-performance cars are all electric-powered. Dodds described to us how a brand with limited marketing resources is able to punch above its weight thanks to the well-aligned agendas of the various parties involved:

> Let's say I want to get the word out about a new event. If I tell the people that follow my Formula E account, "Great news, we've just signed a deal to race in Tokyo," then I'm, you know, telling hundreds of thousands of people. But if I can get all of the *drivers* to say, "Hey, brilliant news! I've just found out I'm racing in Tokyo next year," then all of a sudden that could be tens of millions of people we're telling. If I then get the *city of Tokyo* to say, "Brilliant news: we're going to be hosting a Formula One race here next year," then that could be hundreds of millions of views. So you go through this layering journey of influencers at different levels creating a multiplying effect.

Because of the nature of his business—which itself operates as a constant promoter and influencer of others—Dodds said its reliance on influencers is very high, and yet the expenditures are very low. "If there's a need to get to a certain audience that we can't access, we would consider hiring influencers. But I can't even think of any we have. The very vast majority of ours will be unpaid."

Short of influencing for free, some content creators take their compensation in kind. Rather than receiving money, they may be paid in the form of free merchandise or services or access to exclusive events. There seem to be no limits to the willingness of people to film and say nice things about resort hotels, for example, in return for a complimentary stay. And even Detert joked that if his favorite brand of hummus came knocking, he personally would agree to work for food.

But increasingly, influencers at all levels stand to benefit from monetization that pays off in cold, hard cash. The folks at Influencer Marketing Hub helpfully provide an estimation tool to find out whether you could be cashing in on whatever influence you wield. Its inputs are simple: type in your number of likes and your number of followers, and it spits out a dollar amount for you to consider charging. In fact, the valuation process is hardly that simple, but at its root, the idea that influence comes down to how many people listen to you times how positively they respond is hard to argue with.

Will anything stand in the way of the advancing army of influencers? Perhaps one big and unsettling enemy looms in the wings. For many, the real threat to their livelihoods may be the growing ranks of AI-created virtual influencers. These already exist, and their work is already being monetized: the *Financial Times* reports that advertisers pay $1,000 per post for a popular AI-generated model to promote their wares.[25] But what about authenticity? One media industry insider told the *FT* reporter that backlash from users did not seem to be a problem: "Influencers themselves have a lot of negative associations related to being fake or superficial, which makes people feel less concerned about the concept of that being replaced with AI or virtual influencers." It will be hard, in other words, for influencers to fend off artificial competition if they are seen as phonies themselves.

Whatever inroads AI might make, Detert sees a trend by which increasingly specialized and niche-oriented influencers build audiences that, if they are a fit for a certain marketer, are the most prospect-intensive target markets imaginable. Individual content creators can easily turn into niche influencers as they curate content around their particular interests and, in doing so, build a personal brand. They can develop intensely engaged audiences of like-minded followers who have similar propensities to buy certain kinds of offerings and who trust the judgment of the content creator. Call them everyday celebrities—individually micro, they may be the biggest thing to come out of social media marketing.

Detert said that these relative nobodies, rather than the Kardashians of the world, are the real engine of the influencer economy—especially when they are activated to work in concert, for example, driving an assigned hashtag to trending status. It's only in the past few years that the penny has dropped and marketers have started to grasp this trend. We're now in "the age of micro," Detert said, when brands can see the better returns from working with "someone who may only have two million followers, but they perfectly fit the audience you're looking for—of, say, DIY people in the Midwest, or moms in Delaware." Small-scale content creators, he added, are people who "want to overdeliver" for the brands that choose to engage with them.

Microinfluencers have in some way existed from the beginnings of the social web. In China, early users of WeChat were presented with an attractive new opportunity in Jizan (collecting likes), allowing them to earn freebies or discounts from businesses, including local restaurants. Jizan was essentially a way to gain a financial reward for acting as a promoter of a business, like any tell-a-friend promotion. In this case, someone would repost a promotional message on their own WeChat page and urge their friends on that platform to click through to the seller's official WeChat account and give it a like. Because that traffic was traceable through their account, they could be credited for the referral. Lily, a nineteen-year-old factory worker from a Chinese inland village, was delighted when an art photo studio ran just such a promotion: "I always wanted to [create my own photos], but I didn't save up enough money for that. Last week they launched a sale, and with 50 Likes on WeChat I was able to have a basic set of art photos for half the price."[26]

At this same level of the super microinfluencer, we now have Elon Musk's X site enabling anyone with a verified account to become a premium subscriber. This status allows them to generate revenue from sheer engagement and ad placements in their feed—or to offer their followers a chance to pay them directly as subscribers, in the same way that Twitch does, and receive exclusive content.[27]

Microinfluencing is an important decentralization and democratization of not only publishing but also marketing—and one that will probably bring about a wave of more-targeted businesses. Now that microsegmented marketing has been made possible by the abundance of consumer data produced by social media, many niche opportunities that were simply not economical to pursue could become so. At manageable cost, new products and business models can be tested, and very focused small businesses can be launched and prove profitable. New markets will be created and served—markets that were previously too small and costly to target. Marketers will find that customer bases of just several thousand people can be profitable once they have the ability to zero in on them with a compelling, targeted message.

And, like Ryan Detert and Neil Waller, some will follow the example of the wisest business minds of the US gold rush years. They will see that the right opportunity for them is to stop panning for gold personally and start selling the provisions needed by the next wave of gold diggers.

. . .

Let's zoom in on an individual influencer to understand how this new like-facilitated world is experienced from their perspective.

In 2018, when TikTok burst onto the American social media scene, Michael Le, a Vietnamese American teenager who loved coming up with his own dance routines to hip-hop music, was perfectly positioned to capitalize on the new platform's model. Le had been seriously working on his dance skills since 2012, starting when he was twelve, and stood out for his talent, creativity, and goofy humor. At that young age, too, he was also very drawn, he told us, to the "OG YouTubers, like Ryan Higa or PewDiePie or KevJumba." He's using a term put into the language by the rapper Ice-T's album *O.G. Original Gangster*. The term *OG* is an adjective people have applied ever since to the old-school practitioners in their own fields—people who broke new ground and continue to be revered for that. "These people that I grew up watching," Le said, "they were making a career out of just them having fun, and I wanted to try it out myself. So in 2015, I started posting covers of my dance classes and my own choreographies and, after a year, got my first spark of success where something went viral. Suddenly on all my platforms, I gained between one hundred thousand and two hundred thousand followers."

Then TikTok arrived, and it was a perfect match of creator and platform. TikTok had been launched by ByteDance under another name in 2016, but it was only after ByteDance's merger with another Chinese social network, Musical.ly, in 2018, that the platform went global. The same merger brought an update to TikTok's algorithm, and for whatever mathematical reason, it loved Le's content. "The first week I hopped on TikTok, I gained a million followers," he recalled. "I was shocked. I was just like, 'What's happening? No idea! But I need to keep doing this.'"

"Three to four months in," he added, "I had ten million followers. Then it was twenty million, and my whole life completely upside-down changed." But how did this sudden fame turn into money in his pocket—and so much of it that he would drop out of school, become the main breadwinner of his family, and move his parents and siblings to Los Angeles? TikTok didn't have a monetization program like YouTube's that would pay Le directly for his high audience engagement. The breakthrough came when Le started hearing from brands that they would like to make him part of their social media promotional campaigns.

Elsewhere, he has told the story of his first major deal with a brand—it was with Bang Energy, a line of energy drinks produced by

Florida-based Vital Pharmaceuticals.[28] Soon he would find himself working with Disney and fashion brands like Prada, Hugo Boss, and Ray-Ban. That's where the money becomes very attractive.

But today Le sees an environment transformed by a recent "massive influx of creators" who are incredibly attuned to the metrics that platforms are using to decide what to boost. "I think social media is the biggest game of adapting," he said. "Whatever the algorithm is leaning toward, creators respond to. Maybe it's, 'Oh, carousels are doing better than video," and they'll change up the way they post their Instagram feed. Anything to get more views." And because the change is constant, with algorithms shifting month to month, the average creator finds it "a really hard thing to keep up with. But it's really, really important for them to constantly stay updated, or else they're going to be left in the dust. You'll see it in the numbers, you'll see it in the views, and that's a creator's lifeblood."

So, Le himself is now working on moving to higher ground and more meaningful connections with his audience. He thinks they— and social media users in general—are ready for a new era in their experience. Le said that across the first decade or so, "people were so addicted to this quote-unquote viral content and its dopamine hit, . . . but then after a while they got tired of it and people started wanting to have more of a slower-paced, more organic kind of experience." He thinks someone following an account today is looking for more authenticity, wishing for a greater sense of community, and expecting higher production value.

It's no sure thing that audiences are ready to commit to a deeper relationship with their influencers—maybe this is just wishful thinking from one influencer who is ready to settle down. But maybe Le is right. And if so, he is well positioned to make that transition. And he has a plan, which he laid out for us in essentially three parts. First, he is shifting to a strong emphasis on building his own brand. He will now only partner with other brands that resonate with his own, ones he truly feels good about promoting, featuring products his audience will see as being just the kinds of things he would honestly be enthusiastic about. This natural resonance, he added, should mean that he retains more creative control over how a message is delivered.

Second, he is gravitating toward models that have him being paid more by the audience he wants to engage than by the marketers who want access to that audience. This will mean focusing on platforms supporting membership or subscriber possibilities and basing higher monetization on engagement metrics (including likes but, to an even greater extent, view duration). As he recently explained to *Forbes*, this means shifting away from microvideo platforms, where, "because it's shorter content, it's harder to get more paying money for those things. On YouTube, if you got 100,000 views, that would equate to maybe a couple of thousand dollars. The same 100,000 views on TikTok—that's maybe four dollars or less. So it's a night-and-day difference."

And third, doing this will require that he invest in taking his quality to the next level. "I'm actually trying to move into YouTube," he told us, "because the longer the people watch me, the more of a fan they would become of me." But longer viewing times will put more emphasis on the technical merits of videos. Le believes that "the bar is going to be set way, way higher" for production value in years to come. Artificial intelligence tools are going to help make that easier for individual content creators but at the same time will make it harder for anyone lacking AI skills to compete. "Once these things are accessible to everyone," Le said, "it's like, Who can utilize these tools the best? And I think that's going to be a real game changer."

. . .

There are additional elements to the new commercial ecosystem that has evolved with social media's transformation of marketing and advertising. We've talked about the emergence of social marketing agencies, then the rise of the customer-experience businesses, and the rise of the influencers and their agents—all fundamentally powered by likes. But there are other new players as well. Who else out there is making money on likes? A fair number of these players are in the business of selling and wrangling the data itself, for various reasons.

First are the big-data brokers. Acxiom, for example, buys data from many sites and combines it to create a more comprehensive view of whatever dynamic or phenomenon its customers would like to analyze. One thing it can provide is a truly comprehensive view of any given consumer—by name and in detail. It's a business that grows as fast as the mountains of data available to it. In 2018, according to *Fast Company*, Acxiom had the ability to report accurately on over three thousand attributes of some 700 million people. Just a year later, those numbers had jumped to ten thousand attributes on 2.5 billion people—something like 40 percent of the world's total adult population.

Connected to the rise of data brokers is the development that the value of likes to social media platforms no longer begins and ends with the advertising revenues they generate. Platforms also monetize their users' likes, along with every other action users take, through data licensing. X/Twitter, for example, treats vast amounts of user information as a commodity available for purchase by other organizations—and recently doubled its annual revenue by doing so. These revenues still don't approach those of ad sales, contributing 10 or 11 percent to Twitter's bottom line, but they added up to $571.8 million in 2021 and continue to grow as a proportion of total revenues.[29]

Other kinds of B2B customers, too, have found ways to take the social media platforms' firehoses of data and produce valuable insights for their own customers. These include new media performance-rating agencies, the modern equivalents of Nielsen. New cross-platform marketing trackers like Hootsuite give brand managers a complete dashboard of their social media investments and returns. New media-buying services execute marketers' budget allocation plans through automated bids in ad placement auctions. Put all these entities together, and the like button has made many people a lot of money, not least through individual incomes from jobs that simply did not exist before the age of social media. And on that front, we can add the huge workforces of the platforms themselves, now well stocked with specialized designers and data scientists tending to interface design, programming, user safety, ad sales, user experience analytics, and more.

. . .

In this segment of our journey to make sense of the incredible rise of the like button, we focused on the question of how this small step in UI design turned into big business. We found out some surprising things.

As we trace the story here, we saw likes first being translated to actual dollars for platforms attracting the venture capital investments who foresaw the value of digital engagement. The big payoff came when the social media platforms shifted to ad sales. Today, Facebook, YouTube, Snapchat, Instagram, TikTok, WhatsApp, and others are all more than happy to provide the infrastructure to offer a "free" service to users. Even though the vast majority of their customers don't pay a cent, there are enough message-promoting entities trying to get their content in front of these people that social media turns out to be a very lucrative business.

We also explored how, as the platforms pivoted to ad-sales business models, likes enabled better targeting of advertisements and customer communications (and therefore higher sales). The like button produced new revenues for the brands whose ad dollars were being spent more effectively. Then came a new development with the rise of content creators—small-scale and even one-person purveyors of content that proved popular (liked) enough to be monetized.

But if social media platform owners, marketers and other ad buyers, and influencers are the three most visible creators and harvesters of value in the business of likes, they are hardly the only ones. The like button is also a rainmaker for a whole new ecosystem of businesses that have sprung up or added new capabilities to respond to the needs of all of the above. These new businesses include the niche social media agencies and web analytics vendors serving marketers; the agents, stylists, and production professionals serving influencers; and the data brokers harvesting data to produce insights for other organizations. Finally, the growth of this new commercial ecosystem has also meant new jobs and higher incomes for millions of individuals, from sentiment analysts to influencers, now doing work that did not exist before. In short, money drops from the like button tree in every patch of the social media landscape.

The common theme across all this new value creation is the value of measurable feedback. With liking statistics reliably gauging people's engagement levels and emotional responses, marketers can now largely avoid John Wanamaker's problem. They know which parts of their digital marketing mix are driving revenues beyond their cost. Content creators have the evidence they need to charge appropriately for their influence. And social media platforms have all the data they need to keep refining their algorithms and constructing more-detailed profiles of account holders—and making the whole whirling system of likes and money spin off money even faster.

Of course, wherever there's money, there's always the potential for abuse. The next chapter examines the unanticipated downsides of the burgeoning social media industry.

Chapter 7

Unintended Consequences

n 2019, regulators came for the like button. The UK Information Commissioner's Office—the entity responsible for enforcing the country's General Data Protection Regulation—fingered it as one of those devious "reward loops or positive reinforcement techniques" that "nudge or encourage users to stay actively engaged with a service, allowing the online service to collect more personal data."[1] It had to be reined in. But how could this be? A single technical feature among many, rooted in human sociality and embraced by millions. Why would anyone want to crack down on it?

The problem was that there were multiple unintended consequences. Of course, the impact of the social media industry cannot be entirely attributed to the like button, but we have seen that the button was without doubt a pivotal innovation that enabled social media's rise. These unintended outcomes created pressure on policy makers to respond with new rules to protect the public. It's still unclear how all this will play out, but an unmistakable crescendo of public pressure, regulatory momentum, and the prospect of some sort of reckoning continues to rise.

The like button is hardly unique in this respect. As the original programmer of Facebook's like button, Justin Rosenstein, has explained, "It is very common for humans to develop things with the best of intentions and for them to have unintended, negative consequences."[2] He's right: all successful innovations have spillover effects, technologies create ripple effects, and the social change introduces new challenges that necessitate further change. And most of these consequences are not foreseen.

As we arrive at this point in our journey, we come not to bury or praise. We won't render a final judgment on whether, when all the goods and all the bads are put into the balance, the invention of the like button turns out to be worth it. We're eager to know how that final reckoning will work out, of course, but it's far too soon to say. The case of the like button is a complicated one, and the jury is still out on how problematic and how beneficial it is. Moreover, as communication technologies continue to evolve, so will the challenges and opportunities. Our focus here is rather at the next level up, asking an important question at a point in human history when transformative technological innovations come thicker and faster. We will explore what the specific experience of the like button has to teach about the phenomenon of unintended consequences more broadly and how society can better respond to these consequences.

. . .

First, why aren't we better at foreseeing consequences? Why don't innovators do more to anticipate and head off bad side effects? One simple answer would be that they aren't obliged to do so. Indeed, the way that businesses are measured and managed does not encourage us to focus on such issues. The famous problem of what economists call externalities is that they lie outside the system being measured and are therefore someone else's problem.

Some might argue that innovators should be obliged to be more concerned with the future, and the main question is how to compel them to proceed with more caution. Rachel Botsman, for example, argues that by referring to impacts as *unintended* and *unanticipated*, we are letting the creators of new technologies off the hook. She prefers the term "unconsidered consequences," she said, "because it puts the responsibility for negative outcomes squarely in the hands of investors and entrepreneurs."[3]

Others counter that innovators are rarely guilty of willful ignorance. The failure to anticipate knock-on effects comes down to two other reasons: First and most important, anticipating side effects is much harder than it sounds. In fact, it's usually impossible. And second, for the vast majority of inventions, it wouldn't even be time well spent.

Why is it so hard to foresee emergent consequences? In a complex system, everything ramifies. It's difficult enough to predict simple, linear cause and effect, but the real world isn't a controlled experiment; it's full of connections and evolving conditions, and every new thing introduced creates a chain of secondary effects. The mix of actions and reactions is dynamic.

This is the message of Lee Ross and Richard Nisbett's classic book *The Person and the Situation*: "Social scientists have been pursuing unrealistic goals of prediction. We may never be able to predict . . . how people in general or particular groups will respond to novelty. Situations are highly complex, and so are people's interpretations of them."[4] Ross and Nisbett are "unapologetic" about their colleagues and their own failure to predict, in part because "the situation in the social sciences is not fundamentally different from the situation in the physical sciences. It has long been recognized that the laws of physics do not allow us to predict with much certainty where any particular leaf from a tree will fall." They doubt that the effects of social interventions "can ever be predicted with precision in highly complex, interactive, nonlinear systems."

Many centuries ago, the philosopher Heraclitus observed that "you cannot step twice into the same rivers; for fresh waters are ever flowing in upon you."[5] Unintended consequences crop up constantly because in a dynamic world, every case is unprecedented. It's hard to extrapolate from past experience what will result from a given action. And for the same reason, when a problem arises after the introduction of some new element in an environment, it is often impossible to draw a straight line of causality between the two. Causation and mechanism are hard to figure out in complex, dynamic systems.

Now add to this complexity the likelihood that in the realm of technology, innovations will be applied for unforeseen *purposes*. As discussed in chapter 2, they become part of the toolkit that Brian Arthur described and change the course of events in ways the most thoughtful inventors could not have envisioned. For example, the laser wasn't invented for any particular purpose at all, yet ended up being used for many applications. In 1960, Theodore Maiman, a scientist at the Hughes Research Laboratories in California, developed the first working laser by flashing a high-power lamp on a ruby rod with silver-coated surfaces. Later, a fellow researcher would recall that when this breakthrough was announced, "scientists and engineers were not really prepared for it. Many people said to me—partly as a joke but also as a challenge—that the laser was 'a solution looking for a problem.'"[6] But the development would open up whole new areas of innovation, and "applications came along quite soon." The same can be said for the dynamo and electric motor invented by Michael Faraday in the 1830s (interestingly, after chemist Humphry Davy had failed). But even inventions that were deliberate solutions to well-defined, practical problems end up going in unexpected directions as they become part of the next Lego box of capabilities.

Innovations go in unexpected directions, too, as they are commercialized in different business models. For example, when the new internet search tool Google was launched in 1999, it was immediately hailed as a significant and valuable improvement over the prior experience of browsing the web. Tech journalist Heather Newman, in a column comparing it to AltaVista, Northern Light, and Yahoo, called it "a wonderful thing to behold." The uptake was rapid, sending the site skyrocketing past the established market leaders. Before the year was out, Google had captured a 60 percent market share and reigned as the new ruler of the search engines.[7]

Some years later, Steven Johnson, the bestselling writer about inventions and their impacts, would comment on the blessings of that breakthrough, which turned out to be mixed. Referring to Google's innovative search approach, he said, "That was cause for celebration on almost every level: Google made the entire Web more useful, for free. But then Google introduced a commercial innovation, selling advertisements tied into the search requests it received, and within a few years, the efficiency of the searches (along with a few other online services like Craigslist) had hollowed out the advertising base of local newspapers around the United States. Almost no one saw that coming, not even the Google founders."[8]

Johnson is of two minds about how Google's accomplishment should go down in history. "You can make the argument—as it happens, I would probably make the argument—that the trade-off was worth it," he mused. Optimistically, he predicted that journalism, having had its business model terminally disrupted by Google, would be reborn in some new and better form, more fit for the digital age: "But certainly there is a case to be made that the rise of Web advertising has been, all told, a negative development for the essential public resource of newspaper journalism." This complicated legacy is not an unusual case, he added: "The same debate rages over just about every technological advance."[9]

So, our main observation is that even the *benefits* of innovations, let alone the negative consequences, are mostly unforeseen. So netnet, it's next to impossible to foresee the indirect effects.

And why is it usually *unnecessary* to put a lot of effort into foreseeing consequences? For one thing, most new innovations fail.

Moreover, unintended consequences manifest themselves gradually over time and can be dealt with more specifically and effectively as they emerge. And if something is launched and leads to net bad outcomes, you don't even need policy makers to quash it. People are motivated to end the problem. Solutions coevolve with the problem. Or people just shun the source of the problem.

Of course, the explosive growth of a new innovation could overwhelm such an evolutionary process. We seldom need to anticipate the implications of something growing exponentially—because so few things do. No doubt, some things do grow explosively. In a memorable scene in *Oppenheimer*, a film based on the true story of the development of the first atomic bomb, J. Robert Oppenheimer was asked to assign a probability to the whole world being destroyed by a chain reaction after its first detonation. On that question, for sure, his team had rightly done the math. Similarly, if a team of virologists were working to alter the genotype of a known pathogenic virus, the world would surely demand extreme caution. But the kind of exponential expansion that follows the splitting of atoms (or the spitting of coronaviruses) is the exception and not the rule. Only a tiny fraction of inventions take the world by storm.

Few inventors think too much about, What if this takes the world by storm? The inventors of the automobile didn't envision the ramifications of every household having one. Such a level of success was beyond their dreams. No one planned early or earnestly for the traffic jams, traffic accidents, highway systems, urban sprawl, air pollution, or climate change that would follow. The inventors of personal electronic devices didn't imagine the mountains of e-waste that would accumulate once every person on planet Earth not only had to have a device but wanted to frequently upgrade to the latest one.

Should these innovators have foreseen these ramifications? It's a tough question. It probably does not serve society well to have inventors sit on their hands out of fear that they might unleash something too popular. Too much conscience, to quote Shakespeare, can "make cowards of us all" to the point that "the native hue of resolution / Is sicklied o'er with the pale cast of thought, / And enterprises of great pith and moment . . . / . . . lose the name of action."

. . .

The great sociologist Robert Merton analyzed the problem of "unanticipated consequences of purposive action," mainly in the realm of social policy.[10] His question, too, was why we hadn't as a society become more adept at foreseeing them and heading them off. His great contribution was to approach the subject systematically and, first, to classify unintended consequences into three types: serendipitous benefits, negative side effects, and truly perverse results (in which the very thing you are trying to solve is actually made worse by the intervention). Second, he sorted out five basic causes of unanticipated consequences: ignorance, error, immediate interests overwhelming considerations of longer-term ones, human values preventing actions that would produce better consequences; and the "self-defeating prophecy" attitude that can keep people from deviating from a negative but essentially voluntary course.

All of Merton's analysis was to make the problem of unanticipated consequences more tractable. But just as Ross and Nisbett were not "apologetic about limits to prediction" because they did "not mean that we cannot effectively intervene to better the lives of individuals, groups, or society as a whole," Merton also seems unapologetic, stressing the opportunity costs of spending a lot of time and energy on "trying to see the future" when that "must necessarily take away from alternative uses of those resources on more productive fronts."[11]

It may simply be a fact of human nature that as soon as we gain some new ability to forecast consequences a bit further down the road, we respond by introducing more unpredictability, making change faster or with more audacious scope. Somehow we seem driven to operate at the limit of our reflexes. When we switch on those high beams on a dark rural route, we may relax for a moment at how they have made our drive easier. But then, we step on the accelerator.

. . .

What about the like button—why weren't its consequences foreseen? The ramifications are not exactly Oppenheimer level, perhaps, but on the other hand, as part of the broader rise of social media, they did hit people where they live. We still don't have definitive science on what caused what outcomes, but the ripple effects appear to be pervasive and profound. It is a perfect case study of what we've been describing: an innovation that was unleashed with no hard thinking about what would happen if it got big and what could go wrong, and that then became big in surprising ways and—by some accounts, at least—produced some widespread negative consequences.

In 2019, an article in the British newspaper the *Telegraph* carried a startling headline: "Like Button Most 'Toxic' Feature on Social Media, Royal Society for Public Health Finds."[12] The research described was an RSPH survey in which over two thousand adults and teens were asked to judge various features and aspects of the platforms they used in terms of their helpfulness or toxicity. The verdict was that the like button was considered most harmful—surpassing even those dreaded push notifications that are the bane of so many users' existence.

Setting aside for the moment whether that was a fair assessment (we'll return to the topic of the feature's side effects in a few pages), the like button also illustrates very well why future consequences are hard to predict. We've already seen that it was invented for narrow purposes, like encouraging user-created content, and then got put to wholly different ones. Along the way, the like button also got embedded into business models that hadn't existed when it was invented.

Many other factors were in play in a Web 2.0 ecosystem undergoing a kind of Cambrian explosion, making for a complex environment. And then other inventions were prompted and enabled because the like button had been put into the toolbox. It's obvious that at the out-set, no one expected it to get big. With the like button, no one even anticipated the unintended *good* consequences, let alone the bad ones.

And the like button might also illustrate why we don't need to pre-dict future consequences. We don't necessarily have to foresee bad ef-fects if, as negative consequences emerge, people respond effectively with countermeasures and interventions to reduce them. This is clearly what we have seen and continued to see happen with social media more broadly. For example, in Germany in the spring of 2016, the Düs-seldorf regional court ruled against Peek & Cloppenburg, an online retailer that had been sued by a consumer protection organization for violating shoppers' privacy rights with its integration of the Facebook like button plug-in on its site.[13] As described in chapter 1, a click on the like button would cause the retailer's page or content to show up on the customer's Facebook profile. What's more, a customer might not click anything, but thanks to the use of cookies, the customer's IP address and browser string (revealing the person's type of browser, operating system, and computer) would be automatically sent to Facebook to be recognized if and when the customer ever visited its site. The outcome of the German case shut down this data sharing not only for Facebook but also for LinkedIn, Twitter, Google+, and every other social media site with buttons transferring data in this way.

The many designers involved with the like button over the years didn't anticipate this consequence of the invention—but that was OK, because once the issue cropped up and was deemed unaccept-able, countermeasures were taken. What may seem like a messy se-ries of situations resulting from a failure or unwillingness to plan ahead may be more convincingly described as a social feedback and learning mechanism (akin to a like button on a grander scale) play-ing out quite effectively.

Even companies that profit from the exploitation of liking activ-ity do take action to mitigate negative unintended consequences once these concerns become apparent, because they need to keep their customers satisfied and want to stay out of court. They all do their own internal research on how different features affect users' experience of their services. In 2019, for example, a research team at Instagram surveyed users of its service. Its memo to Facebook top management summed things up: "When people were asked to recall an experience that induced negative social comparison on Instagram, they were likely to attribute that negative feeling to Like counts." And after that, the company took action.[14]

But the story of the like button suggests that there is a special class of things that we perhaps *can't* be content to just let play out. For some innovations, we can't always backtrack or course-correct as we go along, trusting that rationality will prevail on all sides. Some innovations in this world have two features: First, because they scale very rapidly, they produce their unintended consequences on a very large scale before the danger is spotted. And second, because they are highly gratifying (sometimes even addictive), even as downsides crop up and are clear, they don't convince their users to dump them. In some cases, the negative consequence of the innovation is not so much an unrelated side effect but is, rather, the main intended effect taken to an unacceptable extreme. Call it the problem of too much of a good thing: the very aspect of a product that makes it pleasurable can, when taken to excess, make it lousy. As with alcohol, past a certain point, the headiness gives way to headaches.

For this class of innovations, we cannot let innovators off the hook on unintended consequences. Even given all the challenges of prediction, we must work harder to understand if an innovation in the works could have these qualities: Will it scale fast? Will it be excessively gratifying? We need to monitor for early signs that confirm these fears and take prompt corrective action. And we must be more agile and skillful in how we regulate such innovations.

. . .

Learning from the experience of the like button is important if you think there will be more innovations of this kind. We've described it as kind of an outlier because of its rapid scaling and addictive quality. We believe that such innovations could become more common. Not only does digital connectedness facilitate the rapid spread of innovations with social impact, but our understanding of brain science is creating new marketing approaches that could have unintended downsides.

UI designers are now well equipped with an arsenal of so-called nudge techniques—the devious hidden-persuader-style tricks that worried the UK regulators so much. The British Information Commissioners Office defined these developments as "design features which lead or encourage users to follow the designer's preferred paths in the user's decision making."[15]

Nir Eyal, an expert on these kinds of products, packs in the crowds at his seminars on how to deploy them. When he wrote his instructive book *Hooked: How to Build Habit-Forming Products*, he let his readers know that he didn't write it "for the social media companies and video game makers—they already know these tactics." To be sure, some of the underlying science has now been called into question. One of the biggest academic scandals of the century so far was the revelation that many behavioral science studies produced their seemingly amazing findings with altered data. The whole "nudge" approach has turned out to be less potent than it was initially claimed to be. (We're waiting for the new book to be published on this social science debacle, which might best be titled *Fudge*.)

And then there is the growing knowledge of the brain and its reward centers; one branch of this body of knowledge is focused on *neuromarketing*. Back in 2007, Stanford neuroscientist Brian Knutson admitted to *Advertising Age* that using a brain monitoring tool like an electroencephalogram (EEG) to understand consumer preference was like "standing outside a baseball stadium and listening to the crowd to figure out what happened." Twelve years later, he told *Harvard Business Review*, "I look at how far the science has come in the past 15 years, and I'm astonished. We've come so far, so fast. And I really do feel like we're just scratching the surface."[16] With continued progress, it becomes more possible to exert subconscious influence.

In a recent *Harvard Business Review* article, neuroscience professor Moran Cerf shared his concern about what's happening in neuroscience labs at some major companies: "Already they are hiring neuroscientists from my and others' labs, and yet I and others in academia have very little insight into what they are working on. I'm only half joking when I tell people that the moment a tech company introduces an EEG to connect with their home-assistant device— that's when we should all panic."[17]

We need to learn about unintended consequences, using the experience of the like button, because it is a relatable, well-understood example of a growing class of innovations that will likely produce greater repercussions in the future. Even though future technological leaps are wildly unpredictable, they are absolutely inevitable. When change comes at a natural pace and on a local scale, we are well equipped as humans to deal with it, but the "natural" rate and scope were set some millennia ago, before technology gained the power to so rapidly transform our situations. Today, sweeping change can be introduced in a short period, and humans simply have not evolved to stay on top of a highly dynamic landscape and anticipate what it will bring.

. . .

What are the alleged negative unintended consequences of the like button? It's one thing to fear AI making us redundant. We've been doing that as a society for a hundred years already. But this book is about the friendly little like button. Is the button producing such collateral damage that it should be in regulators' sights? How much of that unintended damage is because of the button itself, and how much is because of other aspects of social media business models? What specifically about the feature in context needs to be altered or contained?

As we said at the outset of this chapter, the jury is still out on much of what social media is blamed for—and we mean that literally. As we write these lines, a series of private lawsuits have been brought against social media platforms in which judgments are pending. In the latest, a thirteen-year-old in New York has filed a lawsuit seeking class-action status against Meta (formerly Facebook), claiming that it deliberately worked to get young people like her hooked on Instagram despite the company's knowledge that the social media platform was exposing them to harm. Meta is accused of implementing features its leaders knew to be addictive, "such as displaying counts of how many 'likes' posts receive," even after it gathered evidence that these could harm users' mental health."[18] The suit seeks $5 billion in damages to be distributed to other users if the suit is certified as a class action.

Part of the reason it's hard to prosecute such as case is the difficulty in parsing out what factors were involved and to what degrees. Even if teen mental health has suffered in the period following the introduction of the like button, it's hard to establish causality. Other factors might have contributed—including other internet developments (e.g., online dating sites, role-play gaming or other gaming, or the new reality of limitless porn) as well as offline phenomena like draconian pandemic lockdowns, economic challenges, and the rising use of synthetic drugs.

But there's no point trying to help the like button dodge all charges. We can't have it both ways: we've given the like button a lot of credit for fueling the rise of social media, so it must share some responsibility for the repercussions. The biggest concerns fall into three categories: mental health/addiction, privacy violations/exploitation, and political polarization.

On the mental health side, social media in general gets much scrutiny, but if we stick to this book's purpose of zeroing in on the like button, the argument is this: the button takes something that people engage in naturally but that causes them anxiety—social comparison—and makes it inescapable. Ironically, Leah Pearlman, who was the product manager on the team that developed Facebook's like button, herself fell prey to the anxiety that she wasn't getting enough likes for her fledgling jewelry design business. She resorted to buying ads to goose those numbers—a move she was ashamed to make and owned up to in an interview some years later. She now leaves checking her feed to her social media manager.[19]

The downside of how uplifting it feels to get likes is how dispiriting it feels not to get them—especially when those numbers are on full display to everyone else in your circles. Thanks to breakthrough work in neuroscience, *social pain* is now really understood to be legitimate pain. One fMRI study concluded that "for the first time in humans, it was demonstrated that an experience of social exclusion activated neural regions typically associated with physical pain distress."[20]

The biggest concern is on young users, whose developing brains are especially focused on social sorting in adolescence. As one study found, "technology-based social comparison and feedback-seeking were associated with depressive symptoms. Popularity and gender served as moderators of this effect, such that the association was particularly strong among females and adolescents low in popularity."[21] As we are writing this chapter, New York City has just announced its own official action, with Mayor Eric Adams declaring the "need to protect our students from harm online, including the growing dangers presented by social media. Companies like TikTok, YouTube, Facebook are fueling a mental health crisis by designing their platform with addictive and dangerous features. We cannot stand by and let big tech monetize our children's privacy and jeopardize their mental health."[22] Specifically, the health commissioner was concurrently issuing an advisory "officially designating social media as a public health crisis hazard in New York City."[23]

Jacqueline Sperling is a psychologist who works with young people at McLean Hospital in the Boston area. She knows that people post things on Instagram and other platforms in an effort to shore up their sense of self-esteem and feeling of belonging to the social groups that matter to them, and these needs are satisfied to the extent that they receive positive feedback. It is not surprising, then, that they would fall into habits of constantly checking for likes. But they are also checking the likes that others are getting and worrying about any shortfall on their own part. Beyond the sheer number of likes, they are puzzling over such questions as "Why didn't *this* person like my post, but this *other* person *did*?" Sperling concludes that "they're searching for validation on the internet that serves as a replacement for meaningful connection they might otherwise make in real life."[24]

A second whole category of harms blamed on the like button has to do with violations of privacy. The argument is this: People's use of the like button makes it possible for their personal proclivities to be aggregated, analyzed, and predicted. They become known to the systems, in some ways more than they know themselves. This familiarity fuels some of the positive benefits of the like button, but it is also a reservoir of insight that is easily exploited, to the individual's detriment.

The biggest concerns are the organizations that buy tracking data from data brokers—third-party companies that have grown up in the digital environment and exist purely to buy and sell data about people. You may think such companies should not be allowed to operate—and indeed they are already restricted across much of Europe—but they do exist, and they have plenty of eager customers.

Of course, like button data is just one part of the data torrent such actors draw on, and as we've seen, a decreasingly important one as other data and metrics, like engagement scores, proliferate. Yet the button is a legitimate target for privacy protection advocates because it is such a clear divulgence of personal feelings. To press it is to send an intentional, clear, explicit positive reaction—and yet most people use it with the sense that it is a private act, maybe not at the level of filling out a ballot in a voting booth but a reaction contained within a circle of friends. And the very fact that people are so familiar with it—that they use it so blithely, without suspicion—makes it all the more shocking for them to learn that others are mining such data to profile them, to market to them, and perhaps even to manipulate their opinions. The like button's sunny complexion makes it, for a privacy advocate, conveniently appalling.

Finally, the like button is sometimes blamed for the unintended consequence of social polarization. Here is the argument: As the like button allows a network to curate your feed, delivering more of what you are likely to engage with to keep you on the platform longer, you can wind up in echo chambers. The content served up to you not only reflects your inclination to hold certain beliefs but also filters out information that might usefully challenge them. As your seemingly inconsequential likes steer you into a subgroup, your own Overton window (the range of ideas acceptable to a person or group) begins to shift on some consequential political matter. You can gradually become radicalized, turned into a zealot. Your like data fuels this process not just through curation but through your ability to see like counts indicating others' views in your in-group. A study led by Joseph Walther of the University of California, Santa Barbara's Center for Information Technology and Society showed how the social approval indicated by a like on a post increases people's belief in the content—even when it is a fictitious news story. Specifically, this research enrolled participants who were willing to divulge their own partisan leanings to the researchers. The participants were instructed to retweet political news stories and then observe how many hearts or thumbs popped up in response to their posts. When the story highlighted bad news about the other political party, a clear pattern emerged: the more likes the retweet garnered, the more the retweeter came to agree with the content.[25]

There is apparently no scientific consensus on how social polarization works, however. When we talked with complex system scientist Fernando P. Santos, he shared his opinion that this fear of polarization requires a nuanced assessment. He pointed to recent research published in the scientific journals *Science* and *Nature*.[26] "The take-home message here is that there are not that many side effects. If you compare with offline dynamics, these algorithms, if you tune them, or if you disconnect them, they don't affect that mature-opinion-formation process." Other scientists we spoke with pointed out that the main culprit for social polarization was the megaphone effect of social media's ability to turn every vociferous activist into a broadcasting company.

Still, it remains a logical worry that as we are segmented into what various online networks have determined are our like-minded groups, we may find ourselves spending more time online and in conversation with those new friends. It's not hard to imagine that we will unwittingly come to inhabit echo chambers, where certain norms and options prevail and where fringe ideas can come to seem moderate, conventional, "how everyone sees the matter." We can also imagine that once our social groups divide, they will evolve independently in directions that will make it hard for them to re-aggregate or behave toward each other with respect. When a team from the *Wall Street Journal* analyzed the workings of TikTok, for example, it found a dynamic by which people are sent quickly down rabbit holes.[27] Liking things online may send us into the equivalent of those separated villages in the mountains in centuries past, whose languages progressed down different paths to the point that, within several generations, their speech became unintelligible to each other.

· · ·

As bad as these side effects of like button use may be—mental health issues, privacy violations, and polarization—it's easy to take for granted and overlook the benefits we have obtained from it. Many of the side effects appear to be the too-much-of-a-good-thing extremes of the most positive aspects of liking (which, by the way, were also unanticipated outcomes of the innovation). In making its assessment, society must consider both the benefits and the costs, and regulators should consider how the benefits can be preserved while minimizing the harms.

Here, for example, is a good twist no one anticipated. As chapter 1 described, the button was designed to spur more posting by ordinary people on websites whose business models centered on user-created content. And it did accomplish that goal spectacularly well. But the unanticipated and ultimately much bigger impact in a media environment consisting of news feeds turned out to be how it gave ordinary people a lightweight way to express what they value and want to see more of—ultimately in all kinds of realms of their lives. And even better, their ability to indicate so easily what they like means that more of that good stuff subsequently comes their way. Organizations able to see individuals' feedback can deliver better service and value.

And how about this other positive side effect: all this liking means a lot of us today are getting much more affirmation on a daily basis than we did before the advent of the like button. Many, many people in the 2020s are constantly receiving little Valentines from people whose opinions they care about. To our minds, this is the biggest impact of the like button on the world. One study, for example, by Matthew Pittman and Brandon Reich, found that posters of images on social media showed increased satisfaction with their lives and general happiness—and interestingly, a decreased sense of loneliness. In part because of the liking interaction that happens so profusely around images, these people had an increased sense of intimacy with their peer groups.[28] This study was published in 2016, so we might speculate on how much this medium must have mattered during the pandemic years of 2020 to 2023, when, between personal precautions and mandated closures of many gathering spaces, people were consigned to more solitary lives for many months.

Here's a third positive and big yet unforeseen effect of liking: people connecting so easily around shared interests that they go on naturally to extend more help and advice to each other. For one of us, Martin, two model examples of this cropped up recently. The first sprang from his musical interests, especially as a player and collector of wind instruments. When one day he decided he wanted to learn the didgeridoo, he stumbled on a complication: tourists to Australia who are not knowledgeable about the instruments buy more of these indigenous instruments than anyone else, so there's a glut of crummy products on the market and even outright scamming about provenance. A social media interest group Martin was part of saved the day. Other members there helped him navigate quickly to an authentic Aboriginal cooperative, he made a call (which turned into a delightful four-hour conversation), and within a week, from halfway around the world, Martin had his authentic didgeridoo. How did his connection to that online interest group begin? With a simple click of a like button. The same is true for Martin's recent experience after purchasing an electric bike. The bike was terrific, but the company that sold it to him? Not so much. After the seller went bust, customers were left in the dark about what was going on and how they might continue to get parts or information. The answers all came through online groups that had connected through their shared interest in the product—a group that had initially formed around likes.

Or take the example of Strava, Bob's chosen tool for keeping himself honest on fitness goals. The app has a version of the like button, as so many applications do these days, and this one allows him and his fitness-oriented friends to give each other kudos for their dedication—and sometimes-creative approaches—to fitting in a workout in the midst of a busy day. Anyone who has tried to keep their New Year's resolutions to get into better physical shape knows how hard it can be to maintain the level of self-discipline required. A friendly like can make all the difference to your motivation—and the same could be said for how it spurs creative work (like getting the next chapter of a book drafted) or any other hard but worthwhile task.

James Hong summed up what is going on here: "If you were going to break down what the like does, it's any form of positive affirmation feedback from others that triggers that dopamine hit in my brain and gives me positive feedback from something, right? So, in the workout challenge, if I say, 'Oh look, I ate healthier today' and

I get a like, that's just a little extra reward that means, next time I go out to eat, I am that much more likely to eat healthy." But, he added, it's "not just because I want another like, but also because of the cognitive dissonance that would otherwise be created. Feeling the like made me feel good about myself, and so, if I next time don't eat healthy, there's a dissonance created between how I saw myself when I got that like and myself today—and that's something I want to avoid."

Getting more of the stuff we like, enjoying more affirmation from others, benefiting from mutual support—exactly because these are such obvious boons of the like button, we need to remember that they weren't the original reasons for its creation. They are all unanticipated consequences. But these basic now-taken-for-granted benefits of the like button are what fueled its amazingly fast and extensive spread across the social media universe—and then the entire online universe, and then increasingly the offline universe as well. It's hard to point to anything that has gone more viral than liking itself—and again, nobody saw this coming.

. . .

How do societies respond to negative unintended consequences? They often react by regulating. When a pattern of harm is perceived, the first impulse is often, "There oughta be a law!" Rules and restriction are created to prevent the undesirable outcome from occurring again. And to be sure, the regulation of commercial entities has successfully stamped out loads of bad behaviors, squashed many unintended consequences, and served society's interests in numerous other ways. Just consider the contributions to food safety, drug safety and efficacy, and transportation safety, to name a few ways in which regulation has made us better off.

But traditional ways of regulating are not without challenges. The first is *regulatory lag*, the unavoidable problem that the relatively slow process of legislative response cannot keep up with the pace of new technologies emerging and evolving. By the time the cost-benefit analysis is done, the details get hammered out, and the new rules take effect, the game has often progressed or changed.

Another perennial problem is capture. Even short of bribery or revolving-door-style influence, the regulators seldom have the expert knowledge necessary to exert control in new technical areas. They rely on the regulated organizations to supply the information.

Another challenge is the compliance burden placed on the regulated entities and the fact that regulations are rarely sunsetted. The burden therefore only accumulates, raising the costs of doing business—costs that are passed along to customers.

Well-intended regulations can cause their own unintended side effects in other ways too. A particularly conspicuous example was the regulatory solution the US government settled on during the Prohibition era when it realized that bootleggers were buying alcohol manufactured for industrial use, processing it to remove its harmful contaminants, and selling it for drinkers' consumption. The government's solution was to force the industrial alcohol makers to add methanol to their product—a poison that couldn't easily be removed by the bootleggers. The logic was that people would stop drinking liquor because of the great risk that it would kill them. The problem was that the drinking didn't stop—and over the next seven years, perhaps ten thousand US citizens died from consuming alcohol poisoned as a result.[29]

It surely doesn't help that regulators aren't always clear on what they are trying to accomplish. We previously noted three broad problems that people have criticized the like button for exacerbating: mental health issues, privacy violations, and social and political polarization. But those are dissimilar ills, requiring different measures that are likely to fall under different departments. The challenge is compounded when innovations are being put to an *evolving, multiple* set of purposes.

These challenges are further compounded by the fact that while many producers of consumer goods and services have international reach, regulatory regimes are specific to their individual geographical locations. The result is a patchwork of different rules in different jurisdictions. Certainly, this is true for social media platforms. The European Union's privacy regulation, or General Data Protection Regulation, went into effect in May 2018. Regulations in other locations, like China and the United States, have different objectives, priorities, and details.

In light of all these problems with government regulation, powerful companies in an industry sometimes opt to self-regulate. This was the impulse behind the Global Network Initiative in 2008, when the top social media players of the time (Google, Microsoft, and Yahoo) drew up rules of the game they would all agree to play by. As Verne Kopytoff noted in 2011, "The initiative is modeled on previous voluntary efforts aimed at eradicating sweatshops in the apparel industry and stopping corruption in the oil, natural gas and mining industries."[30]

If this is how we have gone about regulating innovation in the past, how could we do so more effectively, for potentially socially transformative technologies that scale very rapidly? Recently the BCG Henderson Institute, which Martin leads, hosted a gathering of scientists, regulators, financiers, and public intellectuals to deliberate on the question. The headline answer: by making regulation less monolithic, more collaborative, and more adaptive.

What does it mean to make regulation less monolithic? It means acknowledging that the situation (to hark back to Ross and Nisbett) or the particular field of application matters. One size rarely fits all. Regulators must work industry by industry or situation by situation to understand specific problems and remedies.

This approach in turn necessitates collaboration. Working together, regulators, companies, and interest groups not only can delve into the specifics of different areas of application but can also create the awareness, understanding, and momentum that will drive the effective application of new regulations.

The third principle of improving regulation is making it more adaptive, which is to say, allowing regulations to change in response to learning from accumulating experience. Rather than applying an unchanging set of regulations, we can have a changing set of experiments, principles, and regulations that adapt to changing situations and our unfolding knowledge of them. A good example of this approach is the international protocol for aircraft accident investigations, whereby the lessons learned from each incident update the norms to be complied with. Undoubtedly, this adaptive approach has been a major contributor to making flying one of the safest forms of transportation.

In this sense, the fact that there are different rules in different territories can be seen as a potential feature and not a bug. It looks messy when there are competing regulatory regimes making different rules. But that's the kind of tumult in which evolutionary adaptation works—providing that experiences are pooled and learning is extracted and applied. More generally, the messiness we see in responding to the like button may be part of an unruly but effective process of adapting.

Here is the big point: when things scale rapidly and have rapidly propagating, pervasive, deep effects, as fast-moving technologies do, regulators need to up their game. They can't rely on the same models and mindsets that guided them in regulating tech innovations of the past. Our answer to what it means to raise the game is to make regulation less monolithic, more collaborative, and more adaptive.

. . .

Is regulation the only way for society to respond to unintended consequences? No, and actually much can be accomplished through shifting social norms. This is especially true for the type of problem we mentioned previously, where the negative consequence of an innovation is mainly a matter of degree—a surfeit of the innovation's very aspect that produced its positive consequence. Take the like button's ability to make a user's feed more relevant. It's a great benefit—up to a point. As Sam Hall of Samsara told us, "I love—I'm a massive animal fan, and I love, love, love, love, love dogs. And because I happen to have spent a bit longer time on a dog video, and I've liked something, I'm getting served certain things, and I've found all sorts of communities about different dogs and dog walking, and I love it. But it is a little scary. It could be misused."

In the same way that norms take hold and spread in the *use* of the like button, they can take hold in the efforts to curb *abuse* of it. And in fact, this is what we see people doing in the absence of legal restrictions.

This is social psychologist Jonathan Haidt's argument—that we shouldn't only rely on lawmakers to impose regulations that will in any case likely be outdated by the time they take effect. We can as private citizens accelerate the process of norm setting. It's not that Haidt, by arguing against regulation, is trying to go easy on social media companies: in *The Anxious Generation*, he blames them and other purveyors of digital solutions for taking a terrible toll on people, citing greatly elevated rates of suicidal behavior and feelings of despair after the introduction of smartphones. He just sees it as a problem solved more effectively by the self-policing actions of a better-informed population. Haidt has created a list of norms that could take hold at a local level in the family, the school, the neighborhood, and the friend group.

Norms are often reshaped by new research about public health dangers. Researchers do the data analysis and try to get the word out. Some, like Haidt, are good at packaging the latest discoveries for lay audiences. Adam Alter, for example, is one of many who have focused on the downsides of digital technology use. His book *Irresistible: The Rise of Tech Addiction* urges readers to adopt disciplines to keep them from compulsively checking their phones. He offers data on how the compulsive habit interrupts the flow state in which they do their best work.

There is a role for government here, too. Public health authorities in particular translate science into practical implications for behavior change. They can educate the public through public service announcements and education to establish new norms.

And nongovernmental actors can do the same kind of work. The Partnership for a Drug-Free America came up with "This is your brain on drugs" in 1987. (The same year, it also produced the classic spot with a product-of-the-1960s dad demanding to know where his son got the bright idea to try pot and getting the answer "I learned from you, Dad! I learned it by watching you!") And the National Crime Prevention Council introduced McGruff the Crime Dog, beginning in 1980, and his advice on how we could "take a bite out of crime."

We talked to Mathieu Lefèvre, who in 2016 founded More in Common to understand what divides us in Western societies and what can potentially reunite us. In a survey, the nonprofit asked people all around the world what they thought about many aspects of life. It grouped people according to these core values (versus typical demographic categories) and found seven basic "tribes" based on a hidden infrastructure of values. "The US is seen as a fifty-fifty country, red versus blue," he told us. "But actually, with this more complex lens, we get a more nuanced picture that opens up areas for compromise." But what's keeping that compromise from happening? In part, Lefèvre's report on the "Wings," the organization's term for the people at each of the two furthest ends of the political spectrum, fingers the like button:

> Why do the Wings dominate the conversation? A key reason is that polarization has become a business model. Media executives have realized that they can drive clicks, likes, and views and make money for themselves and their shareholders by providing people with the most strident opinions. This means that the most extreme voices—no matter how outlandish— often get the most airtime. In addition, people with the most

extreme views are often the most certain of their positions. They are willing to argue with anyone and avoid moderating their opinions or conceding points to the other side. All this can make entertaining television and viral social media content. But it is distorting how we see each other, fracturing our society, and adding to distortions in our political system that give undue weight to the most extreme views.[31]

More in Common is also trying to be part of the solution to these issues by testing the effect of different novel interventions, like making opposing groups more aware of each other's views or facilitating interactions and conversations between groups. The organization's optimism about possible solutions to social division is enshrined in its very name—its research consistently shows that while the differences between the beliefs of different social groups are real, the perceptions of those differences are almost always exaggerated, meaning that we have more in common than we generally perceive.

The media does its part to establish norms as well. "It's 10 p.m. Do you know where your children are?" was coined by a Manhattan television executive during the long hot summer of 1967, when young people were involved in high levels of urban crime.[32] The fine people at Encyclopedia Britannica made a film in 1951 telling the sad tale of Marty Malone, a "good boy" who tries marijuana and winds up on heroin and in jail. New York's Poison Control Center created a much-loved jingle of singing pills in 1985, underscoring that prescription medicines are often eaten by kids who think they are candy. We haven't yet seen the public service announcements about social media's dangers, but perhaps they are coming.

. . .

Another alternative to heavy-handed regulation is the creation of better-functioning markets so that market forces can do more of the policing. Along with the education of the public just discussed, transparency and competition are the main keys to better-functioning markets. You could argue that the deregulation movement that began in the 1970s mainly reflected a rise in transparency and, consequently, the lower likelihood that markets would fail.

If fierce competition for consumers' and advertisers' business means that sellers of toxic goods and services get punished in the marketplace, then you don't need as much threat of regulation to get voluntary improvements by businesses. They do it in their own economic interests. And sure enough, we have seen some of this in the realm of social media and with the like button in particular. For example, we have already described how internal research at Instagram in 2019 resulted in making likes less visible. Whereas users had been able to see how many likes any posted content had received, they would now see only the likes garnered by their own content. This experiment was first launched only in selected territories: initially in Canada and later in Australia, Brazil, Ireland, Italy, Japan, and New Zealand. The platform started testing removing likes in the United States in November. "The idea is to depressurize Instagram," CEO Adam Mosseri said. "We're trying to reduce anxiety, we're trying to reduce social comparisons." The platform now gives all users the option to hide or unhide the like count on their own and everyone else's posts.[33]

Similarly, in the fall of 2018, Twitter cofounder and then CEO Jack Dorsey openly raised concerns at a company event about how the like button "incentivizes people to want them to go up," adding that he was "not a fan of the heart-shaped button" and that Twitter would be getting rid of it soon.[34] However, forces within the company were apparently working in the opposite direction. Twitter promptly issued a statement assuring users there was no immediate plan to remove the like button.

The moderating effect of markets only works if there is competition. Social media businesses exhibit so-called network effects: a platform is more useful and economic the higher its number of users. Because network effects encourage a tendency toward extreme

concentration, regulators have also become focused on applying an-
timonopoly policy to these businesses.

Amelia Fletcher is a professor and director of the Centre for Competition Policy at University of East Anglia. She conducts interdis-
ciplinary research on competition policy and regulation, did a long
stint as a nonexecutive director on the UK's Competition and Markets Authority board, and was one of the founding panels that culmi-
nated in the UK's new digital markets legislation. She stressed to us
the increasing-returns dynamics that naturally cause concentration
in the social media platforms—and that are very much reinforced by
data-gathering features such as the like button. As a result, she said,
a few platforms have reached enormous scale and are, in practical
terms, impervious to challenge by new entrants. In reality, you cannot go build your own Twitter just because you don't like what you
see in the existing one. If you need proof of this statement, check the
growth and success of Parler, Truth Social, or other challengers.

If you don't feel pressure from any of these directions, the existence of a lofty mission statement is certainly no guarantee. The only
time you need a mission statement like Google's former "Don't be
evil" motto is when you don't have a market mechanism or regulatory body to keep you good.

. . .

When the first telegraph service opened for business—a thirty-eight-
mile line capable of transmitting messages from Washington, DC, to
Baltimore—its inventor, Samuel Morse, tapped out a first message
that hinted at the prospect of unforeseeable effects: "What hath God
wrought?" We started this chapter with the question of what the like
button could teach about unintended consequences of technological innovation. Our goal was not to judge but rather to learn how
unintended consequences come about and to consider, with the like
button as a reference, how innovators and societies might handle
them better.

This chapter showed how unintended consequences are always afoot. Innovation means change, and change means ripple effects. The answer cannot be mainly about better foresight. Yet the like button also serves as a harbinger of a growing class of innovations that can scale so rapidly that society cannot take a laissez-faire attitude toward them. We can and should use the case of the like button to extract bigger lessons about how unintended consequences develop and what we can do about them.

As this chapter described, unintended consequences always inspire calls for regulation, but regulation as it is traditionally approached might be inadequate to the challenge. A more adaptive, collaborative, and less monolithic rule set would be more responsive and effective. Meanwhile, regulation is not the only means of reining in bad side effects. We need more civil society efforts to set and promote healthy norms, and we need markets that (through transparency and competition) are efficient at pushing solutions toward more societally beneficial norms.

Addressing the problem of complexity and the impossibility of heading off unintended consequences, Ross and Nisbett conclude, "The discovery and description of the sources of such inherent unpredictability, whether in the physical sciences or the behavioral sciences, is hardly a cause for apology. It is an important intellectual contribution with profound theoretical and practical implications."[35] In its own small way, the like button helps make this intellectual contribution. The butterfly emoji may have come along only in 2016, when it was approved as part of Unicode 9.0, reports *Emojipedia*. But the like button's butterfly effect was set in motion earlier. The tiny perturbation of the system that would set off the cascade of consequences that gave rise to today's social media environment was the simple momentary action to create a one-click comment.

This is not the last time we will see such a process play out, and the particular case of the like button tells us plenty about how it works, what we can do about it, and why we shouldn't rely on an overly simplistic cartoon version of the challenge, featuring shortsighted innovators and slothful regulators. The like button's repercussions retell an old story. With high-impact innovation, its ramifications will always reach further than can be anticipated. Some bad will always come along with the good, the new solution will always open up new perils and possibilities, and new public policy may be required to monitor and manage them. As Robert Johnson writes, "Technology helps shape the discursive and material characteristics of cultures. As technologies emerge and are incorporated into a cultural context, they alter not just the immediate activity for which they were designed but also have 'ripple effects' that shape cultures in defining ways."[36] Naturally, that statement brings to mind innovations like the printing press, steam engine, and computer, but the impact is not limited to ones whose transformative influences are so well chronicled. "Some mundane technologies, maybe less visible but no less essential," Johnson writes, "have had equally strong cultural effects and are as influential as any other 'shapers' of cultures."

The like button should teach us that with any consequential innovation, the unforeseen outcomes end up exceeding the intended ones—by far. We've noted that in general, it's not sufficient to work harder to anticipate and contain consequences before moving forward; given the success rate of inventions and their typical diffusion rate, that would be time poorly invested. And more importantly, we've had to admit that it's not possible, anyway. Yet as innovators and societies, all of us must keep trying—we must continue to raise our game in problem detection, piloting, regulatory control, norm setting, and harm mitigation. It may not surprise you, then, to see where our conversation about the like button wound up and where this book will end—with the question of what comes next. That is the topic of the final chapter.

Chapter 8

The Future of Likes

Gary Shteyngart was pausing, rubbing his chin thoughtfully, considering the question he had just been asked but hardly pondering it for the first time. Our question to him: What is the future of the like button? The last time he gave the same topic a lot of thought was in 2007–2008, when he was in the thick of writing a novel called *Super Sad True Love Story*. It's an absurdly creative tale of a mismatched pair of lovers in a dystopian near future highly influenced by the rankings on a powerful social media site called GlobalTeens. Under the spell of this platform and using the "äppärät" devices they clutch in their hands, everyone in the novel constantly enters numbers between 1 and 800 to indicate how much they like everyone and everything else around them. And thus, they all behave in ways calculated not necessarily to enhance their actual happiness but always to increase their scores.

Recheck that date now. *Super Sad True Love Story* was written in 2007–2008. This was before Facebook had added its like button and made one-click peer feedback part of the fabric of so many people's lives.

Like other masters of his craft, Shteyngart imagines the fictional future through a simple process of looking at where we are today and what got us here, tracing existing trends that will continue to play out. And with regard to the like button, the story has continued in a way that has him concerned. He knows that people value this handy substitute "for some kind of feeling of being acknowledged" but worries about their gravitating to digital tools for it "because that's what we used to get from *friendships*." He told us he dreads a future spent more in digital interaction, first, because he so relishes the real world: "I love a good steak and a martini and the other pleasures of the physical very much—and the virtual never comes to the same level. It's a pathetic imitation of real life." But even more, he is dismayed that the virtual now comes *close enough* to real life for many social media users that it is harder and harder to get them to expend the marginally greater effort of interacting with people offline. "They spend all their time on this thing because it's been positioned as a kind of alternative. And for many people, what it does is it slowly destroys their ability to have real lives, and that makes them *seek* the substitute." The result is a downward spiral into that pathetic imitation of life he deplores. And, he said, "the more atomized society gets, the more these things have a larger purchase on people's time."

Talking to a writer highly accomplished in painting near-future scenarios was something we knew we should do as we took up the last question of our journey into the belly of the like button. Shteyngart himself would endorse the practice. He tips his hat to William Gibson, for example, and specifically his 1984 cyberpunk novel *Neuromancer*, which "proved to be incredibly predictive of what life would be like when we committed ourselves to the virtual world."[1] At the same time, Shteyngart admitted that the future will probably not be so much of a "hellhole" as he portrayed it in *Super Sad*. It is, he said, just part of his job as a fiction writer to focus on the most interesting ways things could go horribly wrong.

So, will the like button live on and continue to evolve and spread? Without too much thought, we could see a couple of reasons why it might not. For one reason, the next new habit-transforming thing might come along to make likes obsolete. And another: the actions of regulatory agencies and policy makers might cause platforms to hobble this feature. On the other hand, legal and regulatory interventions could instead bring back new strength to the like button if its use makes people hesitate to use other social media features. In 2024, people in the United Kingdom were put on notice that they could be arrested under British law for sharing objectionable content on social media.[2] According to Stephen Parkinson, the nation's director of public prosecutions, an account holder on a social media platform can violate speech restrictions as publisher or as a "republisher," and if the content passed along is later deemed defamatory or having the potential to incite racial hatred, the republisher can be held liable in the same way as the original poster. "If you retweet that, then you're republishing that, and potentially committing that offense," Parkinson said. "We do have dedicated police officers who are scouring social media to look for this material and then follow up with arrests." For our purposes, it's just an example to illustrate that the like button does not exist in a vacuum. If more active law enforcement along these lines means that people need to pause a moment and ponder the possible consequences of sharing, the added friction in the process might cause many to default to the simpler like button to register reactions. And liking data could gain new respect as a factor to weight more heavily in algorithms.

As we asked more people about the future of the button, we also heard about more complications and possibilities—especially since the feature performs such a range of functions. As we've seen, many people had a hand in inventing the like button, and what they variously designed it to do is at best a small part of all it eventually did.

People today have a wide variety of motives for offering one-click emotional reactions and many ways to do it. The original button may have been designed to facilitate compliments that would motivate people to generate more online content and then was intentionally tweaked to recognize social media users' desire to register varied emotional responses to content that caught their attention. But people's liking behavior also reflects their desires to assert identity, solidify in-group status, regulate conversation, show interest in others they would like to know better, maintain social ties, amplify favored opinions, and much more.

So, were we asking more about the future of a one-click feature in an online user interface? Or were we talking more broadly about future solutions to the social and commercial needs those clicks serve, and how people's emotional reactions might be efficiently registered in the future? Will platforms continue to offer the like button as an all-purpose tool, a kind of Swiss Army knife for holding up one's end of a digital conversation, or will each of the button's functions take its own different future path?

Either way, whatever comes next is likely to insinuate itself gradually instead of replacing the current feature in one fell swoop. One of the great historians of innovation, Hugh Aitken, made this relevant point forty years ago:

> It is often the case, when a radically new and different technological system appears on the horizon, that it is at first judged to be less efficient than the system it eventually replaces. There are two main reasons for this. First, the conventional system has had the benefit of considerable developmental improvement since it was introduced and it is familiar to users. The system that challenges it is imperfect, incomplete, risky, and often disconcerting. The second reason is more subtle: the standards of performance by which the new system is appraised have been worked out in terms of the jobs that the old system has done and the criteria especially relevant to those jobs.[3]

Aiken cites the example of the overshot waterwheel, which was hardly instantly disrupted by the invention of the Newcomen engine in the early eighteenth century. By that point, the waterwheel had been in use and been refined for centuries, and for a good long time, it remained the more efficient power source for conventional uses. Anyone contemplating building a new mill would certainly have opted for a site next to a river and power it the old-fashioned way. The only reason the coal-powered Newcomen engine had any customers at all was its usefulness in some specific situations, like pumping water out of shafts in a mining area where coal was cheap. Its few applications, however, "gave the new technology of steam power the toehold it needed to set out on its own course of refinement and development, a course that eventually made it capable of performing functions (such as overland transportation) that no water wheel could ever perform." The same story, Aiken notes, can be told about many other innovations: "A radically new technology does more than merely perform old functions better; it makes it possible to perform functions that the technology it replaces could not perform at all. It not only solves a problem; it 'oversolves' it. It literally creates its own future. One implication is that ... it has to be appraised not in terms of the old functions, but in terms of the new possibilities that it opens up."

It's a timeless process of disruption that is good to keep in mind as we contemplate the future of the like button. Chances are, the next solutions are already happening in some corner of the landscape, being deployed to perform niche tasks, and not being recognized as having potential to do much more. As the famous line attributed to Gibson goes, "The future is already here—it's just unevenly distributed."

So, while it's impossible to see the future, one way to keep your predictions on firmer ground is to look at what novel technologies and methods are gaining a toehold currently, and imagine what more could come from these. It is an extrapolation of capabilities already known rather than a fantasy freely concocted of what marvels we might imagine could be invented next. In that spirit, we talked with people whose work has to do with today's like buttons and asked for their thoughts about what could be next. They include developers, tech innovators, social media users, venture capitalists and other investors, people working on regulatory policy ideas, and other people making a living in our digitized and hypersocial world.

. . .

One confident, near-term expectation we heard was that people will grow increasingly comfortable with responding to content in more-nuanced ways than a simple thumbs-up—while still valuing tools that relieve them of putting their feelings into words. Recall that the original decision at Yelp was to offer three choices for expressing positivity. The version at FriendFeed and then Facebook shrank these choices down to just one option: a feel-good click on a smiley face or thumbs-up. Once the simple utility of that one-icon button conquered the world, platforms began allowing for more nuance again including in some cases, features that enabled the display of negative reactions (and sometimes rethinking these features and removing them). On WhatsApp in 2024, for example, a long press on a message produces a pop-up menu with six reactions to choose from: thumbs-up, heart, laughing, shocked surprise, crying, and praying hands. Overall, the trend is toward expansion of the palettes.

On this point, one of the biggest interface trends in social media since the introduction of the like button is the rise of the emoji. Although emoji were invented in 1997, meaning they predate not only the like button also but all of social media, they saw their real flourishing on platforms designed for social interaction. Friedrich Nietzsche wrote that the "most intelligible factor in language is not the word itself, but the tone, strength, modulation, tempo with which a sequence of words is spoken—in brief, the music behind the words, the passions behind the music, the person behind these passions: everything, in other words, that cannot be written."[4] In their goofy way, emojis fill in for some of that.

At this writing, the online dictionary Emojipedia, which seems to have the most comprehensive collection of them and their meanings, has 3,782 entries, each a Unicode Standard emoji in current use.[5] Many of these are faces, for the obvious reason that the content they are layering back into a conversational exchange is the emotional expression we would pick up in a face-to-face encounter. When the social media analytics company Meltwater decided to chart the emojis used most frequently across twelve social media platforms in 2023, six of the top ten featured expressive faces (and the others, the heart, fire, praying hands, and sparkle emojis, were also emotional symbols).[6] The most popular emoji, used 243 million times that year on those platforms, was the "crying with laughter" face. Second place went to the (usually jokingly used) exaggerated crying face, and third place went to the ROTFL (rolling on the floor laughing) face. If emojis are the evolutionary successor of the like button, then they seem to be taking it into all kinds of richer emotional territory.

Indeed, by 2014, emojis were already so well established that one social media site that was launched compelled its users to communicate using only these modern-day hieroglyphs. (Even account holders' usernames could not use standard spelling.) Emojli, which turned out to be a rather unserious hobby project of two friends, shut down after a year but not before more than seventy thousand people had opened accounts.[7]

. . .

If continued expansion of emotional range seems like a safe bet on the future of the like button, so does continued expansion of application reach. The button is expected to continue proliferating across digital sites. To the extent that it has not already achieved such a saturation level that every site invites its users to respond with a thumbs-up or an equivalent, the coming years will bring us closer to that.

Kevin Warren, the veteran marketer we talked to from UPS, commented on the phenomenon by which customers who encounter something new and useful in one area of their lives start to expect it in other areas, too. Part of the "very different" world he and his colleagues had to learn to navigate after the advent of social media was that customers' service expectations had evolved and were "not just based on our traditional competitor" and how it was treating them. Customers now looked at how they could interact with companies in completely different lines of business. Their ability to get real-time updates on delivery status when they ordered food from DoorDash or Grubhub, for example, or to get direct responses from real people in airlines if they complained about a service experience on social media suddenly made that level of interactivity conspicuous by its absence elsewhere. "Because they are doing business like this all the time," Warren said, "they don't want friction in the process. They want real-time, they want ease, they want transparency—so, all of a sudden, the bar was a lot higher in general, and that's something we had to raise our game on."

So it makes sense that a popular mechanism for registering a quick emotional response would radiate out across the digital world, not necessarily because it was strictly needed by a given application but because users had become so accustomed to it that it was expected. When Microsoft announced in 2023 that it was adding a like button to its comments feature in Microsoft Word, the product manager described the change as customer-requested: "We heard from many of you that you wanted to be able to quickly and easily convey your reaction to a comment like you can do on Word for the web, Teams, or other social media apps. We're happy to deliver!"[8]

It also stands to reason that the like button won't stop radiating once it hits the furthest reaches of the digital world—it will likely increasingly find its way back into physical-world instantiations, too. You may have noticed this already happening. Take, for example, its use by the world's largest third-party operator of airports, Aena. This company has installed screens at key customer service points in the dozens of airports it manages around the world, inviting passengers to weigh in by choosing between different-colored buttons imprinted with a range of facial expressions. According to Aena, these have been used so far over fifty-one million times to rate security checkpoints, baggage claims, VIP lounges, parking lots—and, perhaps most often, an airport's restrooms (figure 8-1).[9]

Offline applications like this will probably increase in years to come—perhaps to the point that "like stations" become as ubiquitous as billboards along a highway. They will probably not be limited to commercial settings. Governments, too, will more habitually invite citizens to "smash that like button" if they are satisfied with the provision of public services, the condition of public spaces, and maybe even the process of policymaking. Readers from New York

FIGURE 8-1

Screen in an airport terminal asking patrons to "like" its facilities (or not)

Source: Courtesy of HappyOrNot

City may remember that when Ed Koch was mayor, he made a habit of collaring constituents to ask, "How'm I doing?" But they may not know the story of how he arrived at the habit. It began when he was still a congressional representative:

> I was trying to get attention, but it wasn't easy. When I first started, I would say, "Good morning," and people would rush by me, into the subway. A few would say, "Good morning," but that was about it. I don't think they were being rude, just indifferent, distracted. Then one morning, just to vary the routine a little, I said, "I'm Ed Koch, your Congressman. How'm I doing?" And people responded. They actually stopped. Sometimes they told me I was doing lousy, but they always stopped. And they talked to me, and I listened.[10]

Not every politician truly wants this unfiltered input, but Koch's journey from trying to "get attention" to landing on a gimmick that actually generated useful feedback is a fun mirror to the history of the like button. Most important, what he was expending a lot of time and shoe leather to do in the 1970s and 1980s can now be done at scale and constantly in a world reshaped by the like button. Perhaps not surprisingly, in March 2024, the White House set up a web page inviting viewers to "react in real time to President Biden's State of the Union address" using emojis—limiting the repertoire of responses, we should note, to only positively charged thumbs-up, heart, and celebrate icons.[11]

. . .

But if liking is here to stay, perhaps that does not imply that the button mechanism will persist. Some speculate that the future of the like button could involve nothing to deliberately click at all. The video-calling service Zoom, for example, now has a hand-gesture recognition capability that translates a user's actual thumbs-up into a registered like. Think of it as the next frontier of frictionless.

Related to this capability is the fast-growing field of gestural robotics, in which users' hands and arms are outfitted with sensor-embedded gloves and their movements are translated by a system that controls a robot's action to match what is being modeled for it. Picture a person piloting a drone by mimicking the intended direction, pitch, and yaw from a room far away. The inventor of such a setup explains: "By using a small number of and plug-and-play algorithms, the system aims to start reducing the barrier to casual users interacting with robots."[12] With continued development, he says, the gestural "vocabulary" for communicating with a robot assistant could usefully expand, generally making it possible to direct it (or other devices) in a more natural way and in many additional scenarios. A thumbs-up could, in other words, be part of a range of standardized gestures that would only need to be flashed to be registered as a reaction.

Even less hands-on is the potential offered by the neural implants already being tested in humans in clinical settings and commercially. Recently, Elon Musk's neurotechnology company Neuralink achieved a milestone by implanting tiny electrodes inside the motor cortex of a human subject, allowing this paralyzed patient to guide robotics solely through his thoughts to perform tasks, beginning with playing chess and video games. And in May 2004, Precision Neuroscience, founded by one of the eight-person founding team of Neuralink, "claimed the world record for the number of electrodes used to detect a person's thoughts—quadrupling the number used to read signals in Neuralink's implant."[13] When we spoke with Precision Neuroscience cofounder Benjamin Rapoport recently, he said, "It's only a matter of time before humans will be able to interact habitually with computing via thought. The question is just when, and how it will operate." One implication for the future of liking is that it may, one day, be possible to give something online a like via thought, without the need to tap anything. Voluntarily, the sensor-equipped person could glance at a digital like button and think to activate it. But less deliberately, people might only engage with content and still have their responses to it automatically captured.

This scenario may sound like a far-off possibility, but already within the past several years, electrode-wielding scientists have been able to apply stimulation to the ventral striatum region of the human brain and bring about dramatic mood changes in, for example, sufferers of clinical depression. In one study, a young woman was receiving such an experimental treatment while distracting herself by doing a needlework project. "When they found the right spot, it was like the Pillsbury Doughboy when he gets poked in the tummy and has that involuntary giggle. I hadn't really laughed at anything for maybe five years, but I suddenly felt a genuine sense of glee and happiness, and the world went from shades of dark gray to just grinning."[14] An intriguing corollary is that the message could be sent in reverse through an implanted device, allowing a person's emotional response to a stimulus to be communicated directly and in the most honest of terms.

Yet this effect could be accomplished without anything as invasive as neural implants—and indeed is being done already, to some extent, through computer analysis of facial expressions, body movements, text, and spoken language. This realm was introduced as early as 1995, when Rosalind Picard published her classic text *Affective Computing*.[15] Since then, many research studies and a slew of practical applications have brought computers further along in their abilities to detect and respond to human emotions.

If all this starts to sound Orwellian to you, it may be because you have indeed read your Orwell. Here is how, in the classic dystopian novel *1984*, he describes the apparatus used by the "Thought Police" of Oceania's government to monitor its citizens' sentiments:

> Any sound that Winston made, above the level of a very low whisper, would be picked up by the telescreen; moreover, so long as he remained within the field of vision which the metal plaque commanded, he could be seen as well as heard. There

was, of course, no way of knowing whether you were being watched at any given moment. How often, or on what system, the Thought Police plugged in on any individual wire was guesswork.[16]

What Orwell did not quite imagine when he published his novel in 1949 was that in an age of artificial intelligence, this guesswork might be eliminated, because the monitors would take no breaks at all. With facial recognition and other advanced software, advanced algorithms, and cheap computing power, the surveillance could be constant and ubiquitous. Interestingly, the much-feared government agency keeping track of people in this way in *1984* is the Ministry of Love (or Miniluv)—established, presumably, with the promise of spreading positivity throughout society via more constant, communicative connection.

· · ·

This brings us to another important question about the future of the like button: What new and different uses might this valuable aggregate data be good for?

One possibility, suggested by recent trends, is that the liking data will be dialed back as a factor in many algorithmic operations. It's still true that if an algorithm knows your likes, it chooses better what to display to you next, and the more you use the like button, the better targeted the content will be. On TikTok, for example, likes are still a strong shaper of your For You Page—add a like to every antic of a Jack Russell terrier or dorm-room-decorating triumph, and your feed and recommendations will quickly follow suit. But what was once the dominant driver of news feeds, follow recommendations, and more has lately been diluted by other data relating to trackable behaviors that might be better proof of preference—metrics like time spent viewing a piece of content, tone analysis of comments, and observable actions (and transactions) occurring afterward.

In a list of social media trends to watch compiled by Hootsuite for 2024, one prediction reads as follows: "Shares will matter more than likes, comments, or followers." The piece explains:

> Comments, likes, and followers can all be faked. Views and impressions are easily inflated. But there's one engagement signal that's much harder to game: shares. Unlike comments, likes, followers, or views, shares represent actual value. When someone shares your content in Stories, DMs, or off-platform, you know they're willing to vouch for you to their own audience. Don't get us wrong: comments and likes are still valuable engagement signals. But if you design your content to be shared, the likes and comments will still follow. The reverse isn't always true. Platforms like Instagram, TikTok, and X have either made share counts public or are testing doing just that—and if that's not a nod to their importance, we don't know what is.[17]

One analyst who saw the declining emphasis on likes coming was John Freeman, an executive at CFRA Research, an independent provider of financial research and analysis. Several years ago, he was predicting that likes would be de-emphasized, in part because of how easy it was getting for people to game the system. Between their increasing understanding of what type of content garners the most likes and their outright buying of them, clout-chasing social media posters were actually distorting the reality that marketers had been so excited to gauge. Advertisers had figured this out, he said, and the new reality was that "they're not interested in the number of likes."[18]

His second point was that even a true like simply didn't give a marketer enough to go on. Advertisers wanted richer and more-specific insights to get them closer to a segment-of-one ideal. As Freeman put it, they wanted to see "not just likes but strong dislikes and how many of those dislikes came from Pennsylvania." And the third point Freeman raised was that social media platforms were under some pressure to move away from likes just to appease the critics for whom like counts had become a powerful talking point as they tried to raise red flags about the dangers of social media.

Yet we also heard how likes would continue to be vital for capturing and analyzing for patterns. From his experience at UPS, Warren described to us an evolution by which companies moved first from analyzing likes in the interests of "keeping you engaged" to honing their precision in "selling to you," but most recently have recognized like data as the key to "tailoring things to you." He predicts that like data will achieve its highest utility in guiding product and service development and supporting more-personalized service to customers.

David Edelman made a related point. With colleague Richard Winger, he wrote the prescient article "Segment-of-One Marketing" some thirty-five years ago but freely admitted that, just as it's impossible to see the future now, it was impossible then. During our conversation, when we complimented him on the foresight, he pointed out a major development he and his coauthor had completely missed:

> The other side of "segment of one" which we did not appreciate when the article was written was the *content* side. Basically segment of one in the early days was about "I've got something to sell—now I know exactly who to send the message to." But with content management systems—and certainly with generative AI and those new capabilities—I can create thousands of variants of that message, one relevant for Martin, another for David, and for everybody else individually. Content management, which we did not see then, was a massive unlock in the early 2000s.

Dan Gardner is the cofounder and executive chair of an agency named Code and Theory, which he founded with a friend in 2001—back when it was a very modern idea to take a digital-first approach to a business. Today, the firm employs some two thousand people focused on helping clients capitalize on a still rapidly evolving world of technology-led creativity. Across the two full decades of the like button's existence, he has been a close observer of what it can do for brands and where it is headed next. Like all marketers, he has always seen its value in terms of the data it provides to guide more precise advertising and communication investments. And lately he has agreed that other metrics such as viewing times and levels of sharing are more correlated with purchases actually made.

Yet these observations don't mean he believes the like button will become irrelevant to sellers of goods and services. His prediction for the coming years? That like data will take on new importance to feed *prediction-oriented* algorithms. "People talk a lot about personalization in targeting," he told us, "but I think the future of personalization will be *anticipation*. And we will use AI to figure out how to anticipate someone's wants, needs, or desires." To deliver on that, he said, will require extensive understanding of behavioral interactions and the emotions surrounding them. "And if that becomes the truth of the future—and I don't see how it could *not* be the truth," Gardner said, the like will rise in relevance as a leading indicator of action and as the training data for prediction algorithms.

He admitted the idea might summon the image of a *Minority Report* kind of world in which your heightened emotional state is spotted as a reliable precursor before you take a follow-on action. The 2002 film starring Tom Cruise and based on Philip K. Dick's novella of 1956 depicts a society in which an individual's intent to commit a violent crime can be detected in time for police to intervene. And, because, naturally, every rational person in such a society would understand that a crime could not succeed, the only crimes still being committed are crimes of passion, where emotions overwhelm rationality.[19]

There's also the question of whether the data will be as transparent as it has been so far. The value of likes might endure even as the argument for showing like *counts* fades. In June 2024, Elon Musk announced that like counts on X (formerly Twitter) would no longer be visible on an account holder's feed, although they would still be attached to the post itself—making them visible to users who clicked through to check out that post.[20] Instagram's Adam Mosseri had already announced a limited experiment to hide like counts on some posts. The reasons might be different: Mosseri seemed to be reacting to pressures from activists concerned with the mental health of young people, who frequently emphasize the toxicity of social comparison. Musk may more simply have wanted a cleaner design. One safe prediction we can offer: just as the algorithms that were used to turn social media data into decisions have always been tweaked, those adjustments will continue to be made.

. . .

Some of the most intriguing speculations came from the people we talked to about the impact of artificial intelligence on the like button. Max Levchin, for example, sees a new and hugely valuable role for liking data to train AI to arrive at conclusions more in line with those a human decision-maker would make.

It's a well-known quandary in machine learning that a computer presented with a clear reward function will engage in relentless reinforcement learning to improve its performance and maximize that reward—but that this optimization path often leads AI systems to very different outcomes than would result from humans exercising human judgment. To introduce a corrective force, AI developers frequently use what is called reinforcement learning from human feedback (RLHF). Essentially they are putting a human thumb on the scale as the computer arrives at its model by training it on data reflecting real people's actual preferences. But where does that human preference data come from, and how much of it is needed for the input to be valid? So far, this has been the problem with RLHF: it's a costly method if it requires hiring human supervisors and annotators to enter feedback.

And this is the problem that Levchin thinks could be solved by the like button. He views the accumulated resource that today sits in Facebook's hands ("I'm sure they have literally on the order of a trillion likes in their databases") as a godsend to any developer wanting to train an intelligent agent on human preference data. And how big a deal is that? "I would argue that one of the most valuable things Facebook owns is that mountain of liking data," Levchin told us. Indeed, at this inflection point in the development of artificial intelligence, having access to "what content is liked by humans, to use for training of AI models, is probably one of the singularly most valuable things on the internet."

This was a striking observation for us because, as we talked to most people, the predictions mostly came from another angle, describing not how the like button would affect the performance of AI but how AI would change the world of the like button. Already, we heard, AI is being applied to improve social media algorithms. Early in 2024, for example, Facebook experimented with using AI to redesign the algorithm that recommends Reels videos to users. Could it come up with a better weighting of variables to predict which video a user would most like to watch next? The result of this early test showed that it could: applying AI to the task paid off in longer watch times—the performance metric Facebook was hoping to boost.[21] When we asked YouTube cofounder Steve Chen what the future holds for the like button, he said, "I sometimes wonder whether the like button will be needed when AI is sophisticated enough to tell the algorithm with 100 percent accuracy what you want to watch next based on the viewing and sharing patterns themselves. Up until now, the like button has been the simplest way for content platforms to do that, but the end goal is to make it as easy and accurate as possible with whatever data is available." He went on to point out, however, that one reason the like button may always be needed is to handle sharp or temporary changes in viewing needs because of life events or situations. "There are days when I wanna be watching content that's a little bit more relevant to, say, my kids," he said. Chen also explained that the like button may have longevity because of its role in attracting advertisers—the other key group alongside the viewers and creators—because the like acts as the simplest possible hinge to connect those three groups. With one tap, a viewer simultaneously conveys appreciation and feedback directly to the content provider and evidence of engagement and preference to the advertiser.

Another major impact of AI will be its increasing use to generate the content itself that is subject to people's emotional responses. Already, growing amounts of the content—both text and images—being liked by social media users are AI generated. One wonders if the original purpose of the like button—to motivate more users to generate content—will even remain relevant. Would the platforms be just as successful on their own terms if their human users ceased to make the posts at all?

This question, of course, raises the problem of authenticity. During the 2024 Super Bowl half-time show, singer Alicia Keys hit a sour note that was noticed by every attentive listener tuned in to the live event. Yet when the recording of her performance was uploaded to YouTube shortly afterward, that flub had been seamlessly corrected, with no notification that the video had been altered.[22] It's a minor thing (and good for Keys for doing the performance live in the first place), but the sneaky correction raised eyebrows nonetheless. Ironically, she was singing "If I Ain't Got You"—and her fans ended up getting something slightly different from her.

More chilling is the trend that the US Federal Communications Commission (FCC) and its equivalents elsewhere have recently cracked down on: uses of AI to "clone" an individual's voice and effectively put words in their mouth. It sounds like them speaking, but it may not be them—it could be an impostor trying to trick that person's grandfather into paying a ransom or trying to conduct a financial transaction in their name. In January 2024, after an incident of robocalls spoofing President Joe Biden's voice, the FCC issued clear guidance that such impersonation is illegal under the provisions of the Telephone Consumer Protection Act, and warned consumers to be careful. "AI-generated voice cloning and images are already sowing confusion by tricking consumers into thinking scams and frauds are legitimate," said FCC chair Jessica Rosenworcel. "No matter what celebrity or politician you favor, or what your relationship is with your kin when they call for help, it is possible we could all be a target of these faked calls."

Short of fraudulent pretense like this, an AI-filled future of social media might well be populated by seemingly real people who are purely computer-generated. As mentioned in chapter 6, such virtual concoctions are infiltrating the community of online influencers and gaining legions of fans on social media platforms. "Aitana Lopez," for example, regularly posts glimpses of her enviable life as a beautiful Spanish musician and fashionista. When we last checked, her Instagram account was up to 310,000 followers, and she was shilling for hair-care and clothing brands, including Victoria's Secret, at a cost of some $1,000 per post. But someone else must be spending her hard-earned money, because Aitana doesn't really need clothes or food or a place to live. She is the programmed creation of an ad agency—one that started out connecting brands with real human influencers but found that the humans were not always so easy to manage.[23]

In a scenario that not only echoes but goes beyond the premise of the 2013 film *Her*, you can also now buy a subscription that enables you to chat to your heart's content with an on-screen "girlfriend." CarynAI, an AI clone of a real-life online influencer, Caryn Marjorie, who had already gained over a million followers on Snapchat when she decided to team up with an AI company and develop a chatbot. Those who would like to engage in one-to-one conversation with the virtual Caryn pay a dollar per minute, and the chatbot's conversation is generated by OpenAI's GPT4 software, as trained on an archive of content Marjorie had previously published on YouTube.[24]

How could all this AI-generated sociality affect the like economy? We can imagine a scenario in which a large proportion of likes is not awarded to human-created content—and not granted by actual people, either. We could have a digital world overrun by synthesized creators and consumers interacting at lightning speed with each other. Surely if this comes to pass, even in part, there will be new problems to be solved, relating to our needs to know who really is who (or what), and when a seemingly popular post is really worth checking out. Do we want a future in which our *true* likes (and everyone else's) are more transparent and inconcealable? Or do we want to retain (for ourselves but also for others) the ability to dissemble? It seems plausible that we will see new tools developed to provide more transparency and assurance as to whether a like is attached to a real person or just a realistic bot. Different platforms might apply such tools to different degrees.

. . .

We can speculate on who, individually and as groups, might become more powerful with the evolution of liking. Will it be the average social media account holder—the ordinary citizen in a more connected society? Or will it only be the platforms that capture and aggregate the data about people's preferences, opinions, and relationships? At the geopolitical level, will the future of the like button do more for democratic or nondemocratic regimes? Extrapolating from selected missteps in privacy, exploitation, and so forth can lead to a miserably dystopian view of the future of liking. But by the same token, the concerns that cause people to fear the worst can spur them to work toward the best. History is full of predictions of dystopian futures that never materialized, and that is because humans are not helpless in the face of unfolding events. We make course corrections as a standard practice as individuals, companies, and governments.

A more appealing prospect is that the future of liking will bring more and more of the good consequences of lightweight expressions of positivity, as adjustments are made on all levels to double down on what makes people feel good. We'll see more evidence that liking certain kinds of stuff causes more of that stuff to come our way, and we'll see more productive, reciprocated liking among people, who will realize net happiness gains as a result. Individuals will find others who are like-minded—serving their deep-seated preferences for homophily—but also find ways to get the doses of heterophily that keeps life interesting and keeps people with different interests and priorities talking to each other. Likes will help people establish the norms and social capital that make life in large social communities easier and more satisfying but will not suppress deviation from norms to such an extreme that conventional wisdom can never be challenged. It may sound Panglossian, but these are natural effects when people throughout a population gain better ways to express what they like and encourage others to do so.

. . .

We started this book with a casual question—How did the like button come out of nowhere and, within an astonishingly short time, become such a ubiquitous feature of life?—and we allowed that question to take us on a learning journey. The answers we uncovered and the people we consulted raised questions of their own: Who invented this little feature, and what were they trying to accomplish? Why did they land on the thumbs-up as the perfect icon for it? How does the like button work, what does it teach us about technology innovation and commercialization, and what new problems did it create? And finally, where will the evolution of digital tools and societal norms take liking next? In short, we wound up covering far more territory, from early human brains to AI algorithms, than we ever imagined at the outset.

The like button turned out to have a great deal to teach us about human nature and the evolutionary psychology behind the hypersociality that distinguishes our species. It held lessons, too, about the nature of technology innovation and how different the messy history of an invention often is from the neat, purposeful, and linear way its story is told. Studying the impact of the like button gave us a deep appreciation of the power of feedback data and the many ways it can be used—some of them very profitable.

And we come to the end of our tour through the like button knowing that most of its story still lies ahead. We know that people crave compliments—and that not enough of them are dispensed in the world. For some reason, we just don't give out likes in abundance without prompts. One explanation comes from research suggesting that people underestimate the positive impact of bestowing a compliment and are therefore more likely to suppress a compliment that they fear might not be perfectly expressed or received in the spirit intended.[25] Whatever the reason, if a like button reduces the friction, that means a more compliment-filled world.

Timeline of Events in the Development of the Like Button

In this appendix, we chronologically summarize the events that contributed to the evolution of the like button. For each event, we name the key people involved; explain the innovation or the new approach taken; and, finally, credit the source of our knowledge of the event.

1990s Telegard BBS (bulletin board system)

BBS emerged in the late 1970s and gained peak popularity in the 1980s and 1990s.

- **Event:** Various voting features emerged, including voting booth on Telegard BBS

- **Key people:** N/A

- **What was new?** First way to submit votes within a community software interface (discussion groups)

Interview with Max Levchin.

Around 1993: Mosaic

- **Event:** Mosaic added bookmarking functionality

- **Key people:** Marc Andreessen and Eric Bina

- **What was new?** Mosaic's hotlists represented one of the first types of bookmarking, allowing users to save content for future reference (via URLs)

Richard MacManus, "1993: Mosaic Launches and the Web Is Set Free," CyberCultural, December 8, 2021, https://cybercultural.com /p/1993-mosaic-launches-and-the-web-is-set-free/.

1998–1999: TiVo

- **Event:** TiVo introduced a remote control with voting functionality (green thumbs-up, red thumbs-down)

- **Key people:** Paul Newby (director of consumer design) and a team of six designers

- **What was new?** First hardware like button; voting functionality to teach the device what programs to record on its own

Interview with Paul Newby.

October 2000: HOTorNOT

- **Event:** HOTorNOT launched with a rating bar (radio button) at the top of the interface

- **Key people:** James Hong and Jim Young

- **What was new?** HOTorNOT may have been the first to provide a one-click mechanism to react to content using JavaScript, displayed the next profile without reloading the page

Interview with James Hong.

2000: Everything2

- **Event:** Everything2 added its own way for users to rate each other's content

- **Key people:** Nathan Oostendorp

- **What was new?** The XP points system allowed users to rate each other's content using a link saying "I like it." This is the earliest known use of the word *like* for this functionality in a social website context. Everything2 also sent an email to the author, telling them that someone liked their work.

Author research on https:// everything2.com/.

2000: Xanga

- **Event:** By December 2000, Xanga contained a link on each weblog to allow users to give eProps

- **Key people:** Biz Stone, along with Xanga founders John Hiler, Marc Ginsburg, and Dan Huddle

- **What was new?** Xanga used this voting data to surface popular user-generated content for the home page and other explore features, and emailed the creator to notify them that they had received eProps

Interview with Biz Stone.

2002: StumbleUpon

- **Event:** StumbleUpon was a social bookmarking service that allowed users to rate web pages using up/down buttons and served up content based on like graphs

- **Key people:** Founders Garrett Camp, Geoff Smith, Justin LaFrance, and Eric Boyd

- **What was new?** First multireaction option in a browser toolbar; first use of like graph to recommend web pages that a user may like

Author research via Wayback Machine on http://stumbleupon .com/.

2003: Del.icio.us

- **Event:** Social bookmarking site Del.icio.us provided functionality to save other people's content

- **Key people:** Founders Joshua Schachter and Peter Gadjokov

- **What was new?** Allowed users to combine finds of other users in different categories on their own set of bookmarks

Author research via Wayback Machine on http://delicious.com/.

2003: Google

- **Event:** Following the acquisition of Blogger by Google, Biz Stone published an internal memo at Google proposing the SMURF concept: single multiuse rating and feedback system

- **Key people:** Biz Stone

- **What was new?** Socialization at Google of the power and potential of a like-button-type feature, from the lessons of eProps at Xanga

Interview with Biz Stone.

2004: Surfbook

Lays a claim on the like

- **Event:** Filed like button patent; patent-holding company sued Facebook and lost

- **Key people:** Dutch programmer Joannes Jozef Everardus van der Meer

- **What was new?** First patent on the like button by the late van der Meer on Surfbook, his early social network

BBC staff, "Facebook sued over 'like' button," BBC, February 11, 2013, https://www.bbc.com/news/technology-21411622.

2004: Digg

- **Event:** Digg allowed users to "dig" content from other users

- **Key people:** Founders Kevin Rose and Jay Adelson

- **What was new?** Voting was visible for each piece of content (e.g., "20 diggs"); Digg added a bury option (with thumbs-down) in 2006

WDD staff, "The History and Controversies of Digg," WebDesignerDepot, November 23, 2009, https://www.webdesignerdepot .com/2009/11/the-history-and-controversies-of-digg-com/ and author research on Wayback Machine for https://digg.com/.

May 2005: Yelp

- **Event:** Yelp adds *useful*, *funny*, and *cool* buttons

- **Key people:** Russ Simmons, Bob Goodson

- **What was new?** First one-click instant reaction buttons; first multiple emotional choice to react to user-generated content

Interviews with Russ Simmons, Jeremy Stoppelman, Max Levchin.

December 2005: B3ta
Lays a claim on the like

- **Event:** Voting buttons to curate content; for each piece of content, visible count on how many "I like this" votes were received

- **Key people:** Rob Manuel

- **What was new?** No new functionality, but a claim on the invention

Matt Locke, "How Likes Went Bad," April 25, 2018, https://medium.com/s/a-brief-history-of-attention/how-likes -went-bad-b094ddd07d4.

November 2005: Vimeo

Lays a claim on the like

- **Event:** Video-upload website Vimeo created a button called *Like* for reacting to content

- **Key people:** Developer Andrew Pile

- **What was new?** The "first like button ever clicked," according to *Fortune*; formatted as a button, where previous references using the word *like* were links

Interview with Max Levchin and author research via Wayback Machine on https://vimeo.com/.

2006: Twitter

- **Event:** Invention of the "follow"

- **Key people:** Biz Stone, Evan Williams, Jack Dorsey

- **What was new?** Following was a form of pseudo-liking, and Twitter used this mechanism instead of having a feature akin to the like button

Interview with Biz Stone.

September 2006: Xanga

- **Event:** Blog site Xanga added a module to the left side of each blog to represent "Who gave the eProps"; the module contained a list of profile links

- **What was new?** Xanga enabled any viewer to see who had given the eProps, thereby making it a more social feature

Interview with Biz Stone and team research via Wayback Machine on https://xanga.com/.

October 2007: FriendFeed

- **Event:** FriendFeed created an aggregated news feed of social media posts from a user's various platforms

- **Key people:** Paul Buchheit, Bret Taylor, Ana Muller

- **What was new?** First to pioneer the use of likes as part of a newsfeed construct, both by making the person who sent the like visible and to modify the content that the users receive; button called like (no thumb icon)

- Facebook was developing its *Awesome* button in parallel but did not launch it; by the time FriendFeed was acquired in 2009, Facebook had already gone all-in with the like button

Interview with Paul Buchheit.

Starting from July 2007: Facebook

- **Event:** In his Quora post, Andrew Bosworth (formerly at Facebook) talks about Facebook's internal "Props" project and "awesome" button prototype at a hackathon

Andrew Bosworth, reply posted to "What's the History of the 'Awesome Button,'" Quora, July 13, 2007, https://www.quora.com /Facebook-company/Whats-the-history-of-the-Awesome -Button-that-eventually-became-the-Like-button-on-Facebook.

February 2009: Facebook

- **Event:** Facebook officially rolls out its like button

- **Key people:** Justin Rosenstein, Leah Pearlman, Jared Morgenstern, Andrew Bosworth

- **What was new?** Popularization of the like; using likes as part of the fundamental concept of a graph, to make smarter and faster recommendations on people you might know

Interview with Paul Buchheit and author research via Wayback Machine on https://facebook.com/.

March 2010: YouTube

- **Event:** YouTube abolished the five-star rating system and replaced it with thumbs-up and thumbs-down buttons; all previous ratings are converted into like and dislike percentages

- **Side note:** In September 2009, YouTube came to a realization that a thumbs-up and thumbs-down system would be more effective than its five-star system, since the majority of views were getting five stars

- **Key people:** Shiva Rajaraman (group's product manager) and founders Chad Hurley, Steve Chen, and Jawed Karim

- **What was new?** Long-standing ratings scale (five stars) turned into thumbs-up or thumbs-down as they were deemed more effective

Author research via Wayback Machine on https://youtube.com.

October 2010: Instagram

- **Event:** Instagram launched with liking functionality for photos

- **Key people:** Founders Kevin Systrom and Mike Krieger

- **What was new?** A relentless focus on photos as the content subject to likes

MG Siegler, "Instagram Launches with the Hope of Igniting Communication through Images," TechCrunch, October 6, 2010, https://techcrunch.com/2010/10/06/instagram-launch/.

2015: Facebook

- **Event:** Facebook added multiple reactions alongside the long-standing like

- **Key people:** Mark Zuckerberg, Geoff Teehan (Facebook's product design director), a team of designers

- **What was new?** Popularization of many icon-based emotional reactions; operating at an unprecedented scale

- **Side note:** Article explaining the process of arriving at the reactions

"Facebook Unveils 'Reactions,' a New Way for Users to Respond to Content," SocialMediaToday, October 8, 2015, https://www .socialmediatoday.com/social-networks/adhutchinson/2015-10-09 /facebook-unveils-reactions-new-way-users-respond-content.

2016: Apple iMessage

- **Event:** Apple added quick-response functionality (called tapback) by which users could tap on a message to quickly mark it with one of six icons, including heart, thumbs-up, or thumbs-down

- **Key people:** Apple team

- **What was new?** First time Apple had incorporated the reaction options to messages

"Apple Previews iOS 10, the Biggest iOS Release Ever," Apple, June 13, 2016, https://www.apple.com/gr/newsroom/2016/06 /13Apple-Previews-iOS-10-The-Biggest-iOS-Release-Ever/.

2021: YouTube

- **Event:** YouTube removed count for dislikes to better protect creators from harassment and to reduce dislike attacks

- **Key people:** YouTube team

- **What was new?** Viewers can still see and use the dislike button, but the exact dislike count is not visible to them (is visible only to the creators)

BBC staff, "YouTube removing dislike 'discourages trolls' but 'unhelpful for users,'" BBC, November 12, 2021, https://www.bbc.com /news/newsbeat-59264070.

2022: WhatsApp

- **Event:** Reactions were added to WhatsApp messages

- **Key people:** Mark Zuckerberg and WhatsApp team

- **What was new?** Adoption of reaction functionality by a major social media platform

"WhatsApp Will Now Enable Users to Add Any Emoji They Like as a Reaction," SocialMediaToday, July 11, 2022, https://www .socialmediatoday.com/news/whatsapp-will-now-enable-users -to-add-any-emoji-they-like-as-a-reaction/626992/.

June 2024: X (formerly Twitter)

- **Event:** X (formerly Twitter) made likes private—for privacy reasons, people can no longer see which users have liked others' posts. Users can still see who liked their own posts, and the overall like count is still visible.

- **Key people:** X director of engineering Haofei Wang, Elon Musk

- **What was new?** Allowing people to like content without others seeing that they liked it

Chantelle Lee, "You Can No Longer See Users' 'Likes' on X," Time, June 12, 2024, https://time.com/6988049/x-makes-likes-private/.

Notes

Chapter 1

1. In 2003, the term *Web 2.0* had yet to be coined. It burst on the scene early in 2004, when Tim O'Reilly articulated the principles and features unifying this fresh wave of businesses and launched the Web 2.0 Conference. Tim O'Reilly, "What Is Web 2.0: Design Patterns and Business Models for the Next Generation of Software," O'Reilly Media, September 30, 2005, https://www.oreilly.com /pub/a/web2/archive/what-is-web-20.html.

2. Bob Goodman, personal journal.

3. The Everything2 site was so called because it replaced a maxed-out Everything database created in 1998 by Nathan Oostendorp. By August 2000, when it ran on servers at Slashdot, it reached the point that users submitted nearly ten thousand new write-ups that month alone. The minimalist site remains in existence today, publishing a trickle of contributions.

4. Kevin S. has gone on to write thousands of reviews for Yelp. To start with the earliest ones archived (starting in 2005), go to Kevin S. profile page, Reviews, Yelp, https://www.yelp.co.uk/user_details_reviews_self?userid =Bf87HcPERF9yiSjb2tQBqw&rec_pagestart=1590.

5. "Napkin Valley Virtual Tour," BCG Henderson Institute, https:// theimaginationmachine.org/napkingallery/.

6. Steve Krug, *Don't Make Me Think: A Common Sense Approach to Web Usability* (New York: New Riders, 2000).

7. Russ Simmons recalled that the move to multiple options was because "the culture at the time was really not just about utility. A lot of our top reviewers were motivated by making reviews that were somewhat useful but also very entertaining or sometimes heartfelt. And so I took a stab at the axes of useful, funny, cool, and bad." Bad was removed, however, within a few days because it disturbed the reviewers, whom Simmons recalled as "a sort of proud, sensitive bunch."

8. John Patrick Pullen, "How Vimeo Became Hipster YouTube," *Fortune*, February 23, 2011, https://fortune.com/2011/02/23/how-vimeo-became-hipster -youtube/.

9. Tom Crosthwaite, "Amazon Removes Ability to Comment on Product Reviews," *Acadia*, December 17, 2020, https://acadia.io/amazon-removes-ability -to-comment-on-product-reviews/.

10. Justin Rosenstein, quoted in Steven Levy, *Facebook: The Inside Story* (New York: Blue Rider/Penguin Random House, 2020), 202.

11. Bernardo Montes de Oca, "What Happened to Foursquare? How It Reinvented Itself to Survive," *Slidebean*, October 5, 2021, https://slidebean.com /story/how-foursquare-survives-evolution-slidebean.

12. According to British computer programmer Nick Pelling, he coined the term *gamification* in 2002, when he was shifting from game design to business analysis and "began to wonder whether the kind of games user-interface I had been developing for so long could be used to turbo-charge all manner of transactions and activities on commercial electronic devices—in-flight video, ATM machines, vending machines, mobile phones, etc." Nick Pelling, "The (Short) Prehistory of 'Gamification,'" *Funding Startups* (blog), August 9, 2011, https://nanodome.wordpress.com/2011/08/09/the-short-prehistory-of -gamification/.

13. FriendFeed was acquired by Facebook on August 10, 2009. For reporting on Facebook's close scrutiny of FriendFeed's like button, see M. G. Siegler, "Facebook Clearly Likes FriendFeed's 'Like' Feature," *VentureBeat*, January 11, 2009, https://venturebeat.com/social/facebook-clearly-likes-friendfeeds-like -feature/; Eric Eldon, "Facebook Gives FriendFeed's 'Like' Button a Thumbs-Up," *VentureBeat*, February 9, 2009, https://venturebeat.com/business/facebook -gives-friendfeeds-like-button-a-thumbs-up/; Jason Kincaid, "Facebook Activates 'Like' Button; FriendFeed Tires of Sincere Flattery," *TechCrunch*, February 9, 2009, https://techcrunch.com/2009/02/09/facebook-activates-like -button-friendfeed-tires-of-sincere-flattery/.

14. Levy, *Facebook*, 202–205.

15. Andrew Bosworth, reply posted to "What's the History of the 'Awesome Button,'" *Quora*, July 13, 2007, https://www.quora.com/Facebook-company /Whats-the-history-of-the-Awesome-Button-that-eventually-became-the-Like -button-on-Facebook; Tom Whitnah, quoted in Alexia Tsotsis, "Facebook's 'Like' Button Used to Be the 'Awesome' Button," *TechCrunch*, October 5, 2010, https:// techcrunch.com/2010/10/05/awesome-this-post/.

16. Helen M. Overland, "What Is Facebook Open Graph?," *Search Engine People* (blog), June 17, 2010, https://www.searchenginepeople.com/blog/what-is -facebook-open-graph.html.

17. Naturally, Twitter announced the change with a tweet ("You can say a lot with a heart. Introducing a new way to show how you feel on Twitter") and included a link to a press release on the company's blog. "Hearts on Twitter," Twitter blog, November 3, 2015, https://blog.twitter.com/en_us/a/2015/hearts -on-twitter.

18. Sarah Perez, "X Weighs Adding a Downvote Button to Replies—But It Doesn't Want to Emulate Reddit," *TechCrunch*, July 2, 2024, https://techcrunch .com/2024/07/02/x-weighs-adding-a-downvote-button-to-replies-but-it-doesnt -want-to-emulate-reddit/.

19. *Good Night Oppy*, directed by Ryan White, written by Helen Kearns and Ryan White, Amazon Studios, 2022. For one account of the film's making, see Brandon Yu, "This Mars Documentary Required Many Sols," *New York Times*, November 23, 2022.

20. Rosa Golijan, "Deceased Programmer Invented 'Like' Button before Facebook, Says Lawyer," *NBC News*, February 8, 2013, https://www.nbcnews .com/tech/tech-news/deceased-programmer-invented-button-facebook-says -lawyer-flna1b8302021.

Chapter 2

1. Jess Joho, "HOTorNOT Shaped the Social Web as We Know It," *Mashable*, 2020, https://mashable.com/feature/hotornot-history-20-year-anniversary.

2. Luis von Ahn, Manuel Blum, Nicholas J. Hopper, and John Langford, "CAPTCHA: Using Hard AI Problems for Security," in *Advances in Cryptology—EUROCRYPT 2003, International Conference on the Theory and Applications of Cryptographic Techniques*, vol. 2656, Lecture Notes in Computer Science (Berlin, Heidelberg: Springer, 2003), 294–311, doi:10.1007/3-540-39200-9_18.

3. John Meurig Thomas, "Sir Humphry Davy and the Coal Miners of the World: A Commentary on Davy (1816) 'An Account of an Invention for Giving Light in Explosive Mixtures of Fire-Damp in Coal Mines,'" *Philosophical Transactions of the Royal Society A* 373 (2015): 20140288, http://dx.doi .org/10.1098/rsta.2014.0288.

4. Thomas, "Sir Humphry Davy."

5. Will Oremus, "How Facebook Designed the Like Button—And Made Social Media into a Popularity Contest," *Fast Company*, November 15, 2022, https://www.fastcompany.com/90780140/the-inside-story-of-how-facebook -designed-the-like-button-and-made-social-media-into-a-popularity-contest.

6. Oremus, "How Facebook Designed the Like Button."

7. Matt G. Southern, "Instagram Enables Likes on Stories," *Search Engine Journal*, February 14, 2022, https://www.searchenginejournal.com/instagram -enables-likes-on-stories/438019/. Note, too, that Instagram decided not to allow likes on a story to be visible to the public but only to allow the creator of that story to see them. This decision aligns with Instagram's other removals of the like button to limit the mental health effects for people whose content does not garner likes.

8. W. Brian Arthur, "Increasing Returns and the New World of Business," *Harvard Business Review*, July 1996, 100–109.

9. Claude Lévi-Strauss, *The Savage Mind* (Chicago: University of Chicago Press) 1968.

10. W. Brian Arthur, Marc Andreessen, and Sonal Chokshi, "Network Effects, Origin Stories, and the Evolution of Tech," *A16Z*, podcast, episode 381, 2012 https://a16z.com/podcast/a16z-podcast-network-effects-origin-stories-and-the -evolution-of-tech/.

11. Thomas Thwaites, *The Toaster Project: Or a Heroic Attempt to Build a Simple Electric Appliance from Scratch* (New York: Princeton Architectural, 2011).

12. W. Brian Arthur, "The Structure of Invention," *Research Policy* 36, no. 2 (March 2007), 274–287.

13. Benge Ambrogi, quoted in Dan Kois, "This Is Going to Change the World," *Slate*, August 1, 2021, https://slate.com/human-interest/2021/08/dean-kamen -viral-mystery-invention-2001.html.

14. Arthur, Andreessen, and Chokshi, "Network Effects."

15. Arthur, Andreessen, and Chokshi, "Network Effects."

16. Magdalena Skipper, quoted in Alice Hazelton, "Why Is Scientific Collaboration Key? 4 Experts Explain," Panel Summary, World Economic Forum, June 8, 2021, https://www.weforum.org/agenda/2021/06/4-views-on-why -scientific-collaboration-is-key-for-the-future/. When Jonathan Adams did the underlying analysis in 2012, he concluded: "Changes in the balance of research

done by the lone scientist and that done by teams can be seen in co-authorship data. Coauthorship has been increasing inexorably. Recently it has exploded." Jonathan Adams, "Rise of Research Networks," *Nature* 490 (October 18, 2012), 335–336.

17. Nobel Prize Outreach, "Full Text of Alfred Nobel's Will," NobelPrize.org., 2024, https://www.nobelprize.org/alfred-nobel/full-text-of-alfred-nobels-will-2/.

18. The Royal Institution, "How to Maximise Your Imagination—with Martin Reeves," YouTube, uploaded by The Royal Institution on April 6, 2023, https://www.youtube.com/watch?v=Z-ChwkxFMg4.

19. For Venter's own account, see "J. Craig Venter on Decision to Leave NIH," CSH Oral History Collection, Cold Spring Harbor Laboratory Digital Archives, March 2, 2006, https://library.cshl.edu/oralhistory/interview/genome-research/challenges-hgp/venter-decision-to-leave-nih/.

20. Philip G. Hubert Jr., *Inventors*, Men of Achievement series (New York: Charles Scribner's Sons, 1896), 99.

21. For a short, accessible account of one such study, see Paul Zak, "Why Your Brain Loves Good Storytelling," hbr.org, October 28, 2014, https://hbr.org/2014/10/why-your-brain-loves-good-storytelling.

22. David Owen, "The Inventor's Dilemma," *New Yorker*, May 17, 2010, 42–50.

Chapter 3

1. César, quoted in "Conversation autour d'un pouce," *Lettres Françaises*, December 30, 1965.

2. *Gladiator*, directed by Ridley Scott (2000; Los Angeles: DreamWorks Pictures).

3. Tim Miller, quoted in Gina Carbon, "Linda Hamilton and Tim Miller Have Blunt Thoughts about *Terminator 2's* Thumbs Up Scene," *Cinemablend*, October 27, 2019, https://www.cinemablend.com/news/2483249/linda-hamilton-and-tim-miller-have-blunt-thoughts-about-terminator-2s-thumbs-up-scene.

4. In fact, Siskel and Ebert even trademarked the phrase "two thumbs up" because, Ebert recalls, "we felt that it was valuable and that we had coined it, originated it, and we didn't want everybody to be using it." Roger Ebert, "The Interviews: 25 Years: Roger Ebert," Television Academy Foundation, interview conducted November 2, 2005, https://interviews.televisionacademy.com/interviews/roger-ebert. They may have had the ambiguity of Latin to thank for this. Horace, *Epistle* (1.18.66) speaks of a sporting achievement being commended "with both thumbs," but it isn't clear how they are gesturing.

5. For the actor's own account, see Henry Winkler, "Henry Winkler on the Genesis of Fonzie's 'Whoa' and the 'Thumbs Up,'" Television Academy Foundation, interview conducted November 10, 2006, https://interviews.televisionacademy.com/interviews/henry-winkler?clip=12054#highlight-clips.

6. Ania Szremski, "James Dean Used to Ride These Motorcycles," *HotCars*, October 20, 2020, https://www.hotcars.com/james-dean-used-to-ride-this-motorcycle/.

7. For that matter, the bikers of the early 1950s were harking back to the *First* World War and the latter part of the Border War against Mexican revolutionaries. As early as 1916, William Harley (later of Harley-Davidson) had designed motorcycles equipped with machine guns, which gave the US Army a distinct advantage over a foe still relying on horses.

8. Arthur Guy Empey, *Over the Top: By an American Soldier Who Went* (1917; reprint Kessinger Publishing, 2010).

9. Sid White, "Chinese Give Thumbs-Up Sign as They Await Evacuation," *Lubbock (TX) Evening Journal*, February 8, 1955, 8.

10. "Thomas De Witt Talmage," *Altoona (PA) Mirror*, April 24, 1937, 8. The article continues: "He achieved worldwide fame. He drew immense congregations, while for thirty years his sermons were printed weekly in religious and secular papers, and in 1901 it was estimated that they were published in 3,600 newspapers in various languages."

11. "Noted Divine Dead," *Indianapolis Journal*, April 13, 1902, 1, 5.

12. "Heaven Looking On," *Peninsula Enterprise*, November 2, 1985, 15, 18.

13. "Talmage Sermon," *Peninsula Enterprise*, November 5, 1904, 24, 16.

14. Stewart paid 80,000 francs, per the *Indiana State Sentinel*, April 19, 1876, 18. Goupil & Cie and Boussod, Valadon & Co. Records, 1846-1919, Book 6, entry no. 6837 (p. 81), Getty Research Institute, Los Angeles, online at https://rosettaapp.getty.edu/delivery/DeliveryManagerServlet?dps_pid=FL1676744. For an illuminating discussion, see DeCourcy E. McIntosh, "Goupil and the American Triumph of Jean-Léon Gérôme," in *Gérôme & Goupil: Art and Enterprise*, ed. Régine Bigorne, trans. Isabel Ollivier: 31-44, 38, https://archive.org/details/geromegoupilarte0000unse/page/38/mode/2up. A few months later, the painting was displayed at the New York Academy of Design and thronged by visitors. After Stewart's death, *Pollice Verso* was sold to another wealthy American, Alfred Corning Clark, and later gifted to the Phoenix Art Museum, where it remains on public display today.

15. Jeremy Maas, *Gambart: Prince of the Victorian Art World* (London: Barrie & Jenkins, 1975).

16. Emily Beeny, "Blood Spectacle: Gerome in the Arena," in *Reconsidering Gerome*, ed. Scott Allan and Mary Morton (Los Angeles: Getty Publications, 2010), 40–53.

17. Lucy H. Hooper, "Léon Gérôme," *The Art Journal* 3 (1877): 26–28, quote on 26, https://doi.org/10.2307/20569011.

18. Zola continued his disdain for Gérôme, saying, "There is barely a parlor in the provinces which does not have an engraving of *Duel After the Ball* or *Louis XIV and Moliere*; in the young men's rooms one finds *The Almeh* and *Phyrne before the Areopagus*; these are titillating subjects reserved for male eyes. More serious people have *Ave, Caesar* or *Death of Caesar*. Monsieur Gérôme works for all tastes." Emile Zola, *Ecrits sur l'art* (Paris: Gallimard, 1991), 374.

19. Amy Lifson, "Ben-Hur: The Book That Shook the World," *Humanities*, November–December 2009, https://www.neh.gov/humanities/2009/novemberdecember/feature/ben-hur-the-book-shook-the-world.

20. Juvenal, *Satires* 3.36.

21. The quote is from "De Spe" ("Of Hope"), authored by an unknown Roman poet. Translation from the Latin of *Anthologia latina sive poesis latinae supplementum*, eds. Franz Buecheler, Alexander Riese, and Ernst Lommatzsch (Leipzig: B.G. Teubner, 1868), 269.

22. Edwin Post, "Pollice Verso," *American Journal of Philology* 13, no. 2 (1892): 213–225.

23. Charlton T. Lewis and Charles Short, *A Latin Dictionary* (New York: Harper & Brothers, 1880).

24. M. Auguste Pelet, "Essai sur un bas-relief découvert en 1845, dans le territoire de Cavillargues, près Bagnols (Gard)" [Essay on a bas-relief discovered in 1845, in the territory of Cavillargues, Near Bagnols (Gard)], in *Mémoires de l'Académie du Gard 1850–1851* (Nimes, France: Durand-Belle, Imprimeur de l'Académie, 1851), 35–41, https://communication.academiedenimes.org/wp-content/uploads/2016/08/M%C3%A9moires-1850-51web.pdf.

25. The omission is especially conspicuous in contrast to today's most prominent rhetoricians—politicians seeking election—whose favorite gesture by far is the feel-good thumbs-up. Albert M. Bacon, *A Manual of Gesture, Embracing a Complete System of Notation, Together with the Principles of Interpretation and Selections for Practice* (Chicago: J. C. Buckbee, 1875).

26. Isa Blagden, *The Crown of a Life* (London: Hurst and Blackett, 1869).

Chapter 4

1. "Hannah" is our pseudonym for a subject in a study led by Lauren Sherman, a cognitive neuroscientist. Since earning her PhD, Sherman has worked as a user-experience researcher at Meta. For full details, see Lauren E. Sherman, Leanna M. Hernandez, Patricia M. Greenfield, and Mirella Dapretto, "What the Brain 'Likes': Neural Correlates of Providing Feedback on Social Media," *Social Cognitive and Affective Neuroscience* 13, no. 7 (2018): 699–707, https://doi.org/10.1093/scan/nsy051.

2. For an overview, see Sören Krach, Frieder M. Paulus, Maren Bodden, and Tilo Kircher, "The Rewarding Nature of Social Interactions," *Frontiers in Behavioral Neuroscience* 4, no. 22 (2010).

3. According to analysis by Kepios, there were just over five billion distinct active social media user identities as of January 2024. While this number equates to about 62 percent of the world's population, these include an unknown number of bots, duplicates, and accounts maintained by entities such as government agencies, businesses, and entertainment promoters. Still, we know that two-thirds of the global population now uses the internet to some degree, and social media is the most popular activity online. Simon Kemp, "Digital 2024 Global Overview Report," DataReportal, January 31, 2024, https://datareportal.com/reports/digital-2024-global-overview-report.

4. Nicholas A. Christakis and James H. Fowler, *Connected: The Surprising Power of Our Social Networks and How They Shape Our Lives* (New York: Little, Brown Spark, 2009); Nicholas A. Christakis, *Blueprint: The Evolutionary Origins of a Good Society* (New York: Little, Brown Spark, 2019).

5. Charlotte Canteloup, William Hoppitt, and Erica van de Waal, "Wild Primates Copy Higher-Ranked Individuals in a Social Transmission Experiment," *Nature Communications* 11, article 459 (2020), https://www.nature.com/articles/s41467-019-14209-8.

6. Stephanie Pappas, "What Fueled Humans' Big Brains? Controversial Paper Proposes New Hypothesis," *Live Science*, March 12, 2021, https://www.livescience.com/human-brain-evolution-prey-size.html.

7. Paul F. Lazarsfeld and Robert K. Merton, "Friendship as a Social Process: A Substantive and Methodological Analysis," *Freedom and Control in Modern Society* 18 (1954): 18–66.

8. Miller McPherson, Lynn Smith-Lovin, and James M. Cook, "Birds of a Feather: Homophily in Social Networks," *Annual Review of Sociology* 27 (2001): 415–444.

9. Feng Fu, Martin A. Nowak, Nicholas A. Christakis, and James H. Fowler, "The Evolution of Homophily," *Scientific Reports* 2, no. 845 (2012), doi:10.1038/srep00845.

10. See, for example, Susan Harter, Clare Stocker, and Nancy S. Robinson, "The Perceived Directionality of the Link between Approval and Self-Worth: The Liabilities of a Looking Glass Self-Orientation among Young Adolescents," *Journal of Research on Adolescence* 6 (1996): 285–308; Karen D. Rudolph, "Adolescent Depression," in *Handbook of Depression*, ed. Ian H. Gotlib and Constance L. Hammen, 2nd ed. *(New York: Guilford Press*, 2009), 444–446.

11. Lauren E. Sherman et al., "Social Media 'Likes' Impact Teens' Brains and Behavior," *Association for Psychological Science*, May 31, 2016, https://www.psychologicalscience.org/news/releases/social-media-likes-impact-teens-brains-and-behavior.html.

12. James Olds and Peter Milner, "Positive Reinforcement Produced by Electrical Stimulation of Septal Area and Other Regions of Rat Brain," *Journal of Comparative and Physiological Psychology* 47, no. 6 (1954): 419.

13. David Linden, *The Compass of Pleasure: How Our Brains Make Fatty Foods, Orgasm, Exercise, Marijuana, Generosity, Vodka, Learning, and Gambling Feel So Good* (New York: Viking, 2011).

14. Keise Izuma, Daisuke N. Saito, and Norihiro Sadato, "Processing of Social and Monetary Rewards in the Human Striatum," *Neuron* 58, no. 2 (2008): 284–294, https://doi.org/10.1016/j.neuron.2008.03.020.

15. Andrew Huberman, "Tools to Manage Dopamine and Improve Motivation and Drive," *Neural Network Newsletter*, October 6, 2022, https://www.hubermanlab.com/newsletter/tools-to-manage-dopamine-and-improve-motivation-and-drive.

16. Anna Lembke, *Dopamine Nation: Finding Balance in the Age of Indulgence* (New York: Dutton, 2021).

17. Andrew Huberman, "Dr. Kyle Gillett: Tools for Hormone Optimization in Males," *Huberman Lab*, podcast, episode 102, December 12, 2022, https://www.youtube.com/watch?v=O640yAgq5f8. Huberman devotes a fuller episode later to the topic of dopamine: Andrew Huberman, "Leverage Dopamine to Overcome Procrastination and Optimize Effort," *Huberman Lab*, podcast, March 27, 2023, https://www.hubermanlab.com/episode/leverage-dopamine-to-overcome-procrastination-and-optimize-effort.

Chapter 5

1. Sam Machkovech, "Report: Facebook Helped Advertisers Target Teens Who Feel 'Worthless,'" *Ars Technica*, May 1, 2017, https://arstechnica.com/information-technology/2017/05/facebook-helped-advertisers-target-teens-who-feel-worthless/.

2. Johan Ugander, Brian Karrer, Lars Backstrom, and Cameron Marlow, "The Anatomy of the Facebook Social Graph," working paper, November 18, 2011, https://arxiv.org 1111.4503.pdf.

3. Interestingly, the "six degrees of separation" idea was first noted not in a scientific paper but in a 1929 short story by Hungarian comic writer Frigyes Karinthy. No attempt was made to test its validity till the 1960s, when Stanley Milgram fielded a rather inconclusive study involving the mailing of postcards. In the era of social media, a study using Facebook's social graph—at a point when it had 721 million active users with a total of 69 billion friendship links—found that the number of links between any two randomly chosen users averaged 3.74. Lars Backstrom, Paolo Boldi, Marco Rosa, Johan Ugander, and Sebastiano Vigna, "Four Degrees of Separation," in *Proceedings of the 4th Annual ACM Web Science Conference* (New York: Association for Computing Machinery, 2012), 33–42, https://doi.org/10.1145/2380718.2380723.

4. Facebook/Meta, "Facebook Unveils Platform for Developers of Social Applications," news release, May 24, 2007, https://about.fb.com/news/2007/05 /facebook-unveils-platform-for-developers-of-social-applications/.

5. Alan Zeichick, "How Facebook Works," *Technology Review*, June 23, 2008, https://www.technologyreview.com/2008/06/23/219803/how-facebook-works/.

6. Pamela Vagata and Kevin Wilfong, "Scaling the Facebook Data Warehouse to 300 PB," *Engineering at Meta* (blog), April 10, 2014, https://engineering .fb.com/2014/04/10/core-infra/scaling-the-facebook-data-warehouse-to-300-pb/.

7. For X/Twitter's explanation of its algorithm, see the blog maintained by the company's engineers: "Twitter's Recommendation Algorithm," Twitter / X (2023): https://blog.x.com/engineering/en_us/topics/open-source/2023 /twitter-recommendation-algorithm.

8. Edward Ross, "The Man behind the Curtain: The Algorithms of Social Media," *INKSpire*, May 1, 2018, https://inkspire.org/post/the-man-behind-the -curtain-the-algorithms-of-social-media/-KZd7qog2I8NXTEC5p2.

9. See Matt McGee, "EdgeRank Is Dead: Facebook's News Feed Algorithm Now Has Close to 100K Weight Factors," MarTech, August 16, 2013, https:// martech.org/edgerank-is-dead-facebooks-news-feed-algorithm-now-has-close -to-100k-weight-factors/.

10. Gagliardi, "This Is How the Instagram Algorithm Works in 2024," *Later* (blog), November 29, 2023, https://later.com/blog/how-instagram-algorithm -works/.

11. Gagliardi, "This Is How the Instagram Algorithm Works in 2024."

12. Rainer Maria Rilke, *The Book of Hours: Love Poems to God* (*"Stundenbuch"*) (New York: Riverhead Books, 1996).

13. Emilio Ferrara and Zeyao Yang, "Quantifying the Effect of Sentiment on Information Diffusion in Social Media," *PeerJ Computer Science* 1 (September 30, 2015): e26, doi:10.7717/peerj-cs.26.

14. Mike Isaac, "Facebook Wrestles with the Features It Used to Define Social Networking," *New York Times*, October 10, 2021, https://www.nytimes.com /2021/10/25/technology/facebook-like-share-buttons.html.

Chapter 6

1. Will Oremus, "How Facebook Designed the Like Button and Made Social Media into a Popularity Contest," *Fast Company*, November 15, 2022, https:// www.fastcompany.com/90780140/the-inside-story-of-how-facebook-designed -the-like-button-and-made-social-media-into-a-popularity-contest.

2. Stan Davis and Bill Davidson, *2020 Vision: Turbocharge Your Business Today to Thrive in Tomorrow's Economy* (New York: Simon & Schuster, 1991).

3. Wu Youyoua, Michal Kosinski, and David Stillwell, "Computer-Based Personality Judgments Are More Accurate Than Those Made by Humans," *Proceedings of the National Academy of Sciences* 112, no. 4 (January 27, 2015): 1036–1040, https://www.pnas.org/doi/epdf/10.1073/pnas.1418680112.

4. Michal Kosinski, David Stillwell, and Thore Graepel, "Private Traits and Attributes Are Predictable from Digital Records of Human Behavior," *Proceedings of the National Academy of Sciences* 110, no. 15 (2013): 5802–5805.

5. Richard Winger and David Edelman, "Segment-of-One Marketing," *Perspectives*, Boston Consulting Group, 329 (1989), https://web-assets.bcg.com/img-src/Segment_of_One_Marketing_Jan1989_tcm9-139613.pdf.

6. Joe McCambley, "Stop Selling Ads and Do Something Useful," hbr.org, February 12, 2013.

7. Adam P. Schneider, "Facebook Expands Beyond Harvard," *Harvard Crimson*, March 1, 2004.

8. Meta/Facebook, "Facebook Unveils Facebook Ads," news release, Facebook, November 6, 2007, https://about.fb.com/news/2007/11/facebook-unveils-facebook-ads/.

9. Leslie K. John, Daniel Mochon, Oliver Emrich, and Janet Schwartz, "What's the Value of a Like?," *Harvard Business Review*, March–April 2017.

10. The research was led by Andrew Lipsman of Comscore and Graham Mudd of Facebook. See Carmela Aquino, "Comscore and Facebook Release Research Paper 'The Power of Like 2: How Social Marketing Works,'" Comscore, news release, June 12, 2012, https://www.comscore.com/Insights/Press-Releases/2012/6/comScore-and-Facebook-Release-The-Power-of-Like-2-How-Social-Marketing-Works.

11. Roger Horberry, "The 10 Most Important Social Media Statistics for 2023," GWI blog, https://blog.gwi.com/marketing/social-media-statistics/.

12. Christine Moorman et al., "The CMO Survey Highlights and Insights Report," Deloitte, Duke University Fuqua School of Business, and American Marketing Association, fall 2023, 41, 42, https://cmosurvey.org/wp-content/uploads/2023/09/The_CMO_Survey-Highlights_and_Insights_Report-Fall_2023.pdf.

13. David Ogilvy, *Confessions of an Advertising Man* (New York: Athenum, 1963), 59.

14. Simon Kemp, "Digital 2024 Global Overview Report," DataReportal, January 31, 2024, https://datareportal.com/reports/digital-2024-global-overview-report.

15. McCambley, "Stop Selling Ads and Do Something Useful."

16. Gunwoo Yoon et al., "Facebook Likes and Corporate Revenue: Testing the Consistency between Attitude and Behavior," *International Journal of Advertising* (2024), doi.org/10.1080/02650487.2024.2322855.

17. Alpa Shah, "Data Drives CX Success in 2022, with Social Media Leading the Way," *Customer Contact* (blog), Frost & Sullivan, 2022, https://www.customercontactmindxchange.com/data-drives-cx/.

18. Eric Schmidt, quoted in McCambley, "Stop Selling Ads and Do Something Useful."

19. "The Influencer Report: Engaging Gen-Z and Millennials," Morning Consult, November 2019, https://blog.hostalia.com/wp-content/uploads /2019/11/2019-influencer-report-engaging-gen-z-millennials-morning-consult -informe-blog-hostalia-hosting.pdf.

20. Jeremy Salvucci, "MrBeast's Net Worth: How Much Does YouTube's Top Creator Make?" *The Street*, November 7, 2023, https://www.thestreet.com /personal-finance/mrbeast-net-worth.

21. Statistics published by SpeakRJ (social media auditor). For current performance of influencer MrBeast, SpeakRJ, "MrBeast," https://www.speakrj .com/audit/report/UCX6OQ3DkcsbYNE6H8uQQuVA/youtube.

22. Moorman et al., "The CMO Survey."

23. Brooke Erin Duffy, "Empowerment through Endorsement? Polysemic Meaning in Dove's User-Generated Advertising," *Communication, Culture and Critique* 3, no. 1 (2010): 26–43, https://doi.org/10.1111/j.1753-9137.2009.01056.x.

24. Sara Pudvelis, "Matter Survey Reveals Consumers Find Influencers More Helpful and Trustworthy than Brands during the Pandemic," news release, Matter Communications, May 26, 2020, https://www.businesswire.com /news/home/20200526005058/en/Matter-Survey-Reveals-Consumers-Find -Influencers-More-Helpful-and-Trustworthy-than-Brands-During-the -Pandemic.

25. Christina Criddle, "How AI-Created Fakes Are Taking Business from Online Influencers," *Financial Times*, December 29, 2023, https://www.ft.com /content/e1f83331-ac65-4395-a542-651b7df0d454.

26. Xinyuan Wang, *Social Media in Industrial China* (London: University College London Press, 2016), https://library.oapen.org/bitstream/handle /20.500.12657/32063/1/618832.pdf.

27. X Blue (formerly Twitter Blue) first launched in 2021 as a subscription service offering enhanced features like undoing a tweet and saving bookmarks to folders. Musk relaunched the program in November 2022 and introduced major changes like a blue checkmark in the features for paying users.

28. Forbes, "How Michael Le Shot to TikTok Superstardom and Launched a Web 3.0 Business," YouTube video, posted December 11, 2022, https://www .youtube.com/watch?v=_WUZZRyqA60.

29. Mansoor Iqbal, "Twitter Revenue and Usage Statistics (2023)," *Business of Apps*, August 10, 2023, https://www.businessofapps.com/data/twitter-statistics/.

Chapter 7

1. Stephanie Bodoni, "Facebook's Like Button Makes Websites Liable, Top EU Court Rules," *Bloomberg Businessweek*, July 29, 2019, https://www.bloomberg .com/news/articles/2019-07-29/facebook-s-like-button-makes-websites-liable -top-eu-court-rules?embedded-checkout=true.

2. Justin Rosenstein, quoted in Paul Lewis, "'Our Minds Can Be Hijacked': The Tech Insiders Who Fear a Smartphone Dystopia," *Guardian*, October 6, 2017, https://www.theguardian.com/technology/2017/oct/05/smartphone -addiction-silicon-valley-dystopia.

3. Rachel Botsman, "Tech Leaders Can Do More to Avoid Unintended Consequences," *Wired*, May 24, 2022, https://www.wired.com/story/technology -unintended-consequences/.

4. Lee Ross and Richard E. Nisbett, *The Person and the Situation: Perspectives of Social Psychology* (Philadelphia: Temple University Press, 1991), 17.

5. Heraclitus, Fragment 12, *Fragments of Heraclitus*.

6. Charles H. Townes, "The First Laser," in *A Century of Nature: Twenty-One Discoveries That Changed Science and the World*, ed. Laura Garwin and Tim Lincoln (Chicago: University of Chicago Press, 2003).

7. Heather Newman, "Searching for a New Engine? Size These Up," *Detroit Free Press*, September 26, 1999, 3F. As of January 2024, Google's market share had topped 91 percent.

8. Steven Johnson, *How We Got to Now: Six Innovations That Made the Modern World* (New York: Riverhead, 2014), 7.

9. Johnson, *How We Got to Now.*

10. Robert K. Merton, "The Unanticipated Consequences of Purposive Action," *American Sociological Review* 1, no. 6 (December 1936): 894–904.

11. Ross and Nisbett, *The Person and the Situation*, 18; Merton, "Unanticipated Consequences."

12. Mike Wright, "Like Button Most 'Toxic' Feature on Social Media, Royal Society for Public Health Finds," *Telegraph*, August 30, 2019, https://www .telegraph.co.uk/news/2019/08/29/like-button-toxic-feature-social-media-royal -society-public/.

13. Reuters, "German Court Rules against Use of Facebook 'Like' Button," Reuters, March 9, 2016, https://www.reuters.com/article/technology/german -court-rules-against-use-of-facebook-like-button-idUSKCN0WB24N/.

14. Mike Isaac, "Facebook Debates What to Do with Its Like and Share Buttons," *New York Times*, October 25, 2021, https://www.nytimes.com /2021/10/25/technology/facebook-like-share-buttons.html.

15. UK Information Commissioners Office, "Nudge Techniques," in *Age-Appropriate Design: A Code of Practice for Online Services*, 2022, 72, https://ico .org.uk/for-organisations/uk-gdpr-guidance-and-resources/childrens -information/childrens-code-guidance-and-resources/age-appropriate-design-a -code-of-practice-for-online-services/13-nudge-techniques/.

16. Brian Knutson, quoted in Mya Frazier, "Hidden Persuasion or Junk Science?," *Advertising Age*, September 10, 2007; Eben Harrell, "Neuromarketing: What You Need to Know," hbr.org, January 23, 2019, https://hbr.org/2019/01 /neuromarketing-what-you-need-to-know.

17. Moran Cerf, quoted in Eben Harrell, "Neuromarketing."

18. Naomi Nix, "Teenager Sues Meta Over 'Addictive' Instagram Features," *Washington Post*, August 5, 2024, https://www.washingtonpost.com/technology /2024/08/05/meta-lawsuit-teen-mental-health/.

19. Victor Luckerson, "The Rise of the Like Economy," *Ringer*, February 15, 2017, https://www.theringer.com/2017/2/15/16038024/how-the-like-button -took-over-the-internet-ebe778be2459.

20. Naomi I. Eisenberger, "The Neural Bases of Social Pain: Evidence for Shared Representations with Physical Pain," *Psychosomatic Medicine* 74, no. 2 (2012): 126–135, doi: 10.1097/PSY.0b013e3182464dd1; Sook-Lei Liew, Shihui Han, and Lisa Aziz-Zadeh, "Familiarity Modulates Mirror Neuron and Mentalizing Regions during Intention Understanding," *Human Brain Mapping* 32, no. 11 (2011): 1986–1997, doi: 10.1002/hbm.21164.

21. Jacqueline Nesi and Mitchell J. Prinstein, "Using Social Media for Social Comparison and Feedback-Seeking: Gender and Popularity Moderate Associations with Depressive Symptoms," *Journal of Abnormal Child Psychology* 43 (2015): 1427–1438, doi: 10.1007/s10802-015-0020-0.

22. Miranda Nazzaro, "New York City Mayor Classifying Social Media as 'Public Health Hazard,'" *The Hill*, January 25, 2024, https://thehill.com /policy/technology/4428745-new-york-city-social-media-public-health-hazard -environmental-toxin-mayor-eric-adams/.

23. Kelsey Ables, "New York City Designates Social Media a Public Health Hazard," Washington Post, January 25, 2024, https://www.washingtonpost.com /technology/2024/01/25/nyc-social-media-health-hazard-toxin/.

24. Jacqueline Sperling, quoted in "The Social Dilemma: Social Media and Your Mental Health," McLean Hospital, January 18, 2023, https://www .mcleanhospital.org/essential/it-or-not-social-medias-affecting-your-mental -health.

25. Joseph B. Walther, Zijian Lew, America L. Edwards, and Justice Quick, "The Effect of Social Approval on Perceptions Following Social Media Message Sharing Applied to Fake News," *Journal of Communication* 72, no. 6 (2022): 661–674, https://doi.org/10.1093/joc/jqac033; Shelly Leachman, "'Liking' Is Believing," *The Current* (blog), University of California, Santa Barbara, November 29, 2022, https://news.ucsb.edu/2022/020784/liking -believing.

26. Andrew M. Guess et al., "How Do Social Media Feed Algorithms Affect Attitudes and Behavior in an Election Campaign?," *Science* 381, no. 6656 (2023): 398–404, doi: 10.1126/science.abp9364; Sandra González-Bailón et al., "Asymmetric Ideological Segregation in Exposure to Political News on Facebook," *Science* 381, no. 6656 (2023): 392–398, doi: 10.1126/science.ade7138; Brendan Nyhan et al., "Like-Minded Sources on Facebook Are Prevalent but Not Polarizing," *Nature* 620 (2023): 137–144; Jonathan Vanian, "New Research on Facebook Shows the Algorithm Isn't Entirely to Blame for Political Polarization," CNBC, July 27, 2023, https://www.cnbc.com/2023/07/27/science-and-nature -studies-on-facebook-show-algorithm-not-only-problem.html.

27. WSJ Staff, "Inside TikTok's Algorithm: A WSJ Video Investigation," *Wall Street Journal*, July 21, 2021, https://www.wsj.com/articles/tiktok-algorithm -video-investigation-11626877477.

28. Matthew Pittman and Brandon Reich, "Social Media and Loneliness: Why an Instagram Picture May Be Worth More Than a Thousand Twitter Words," *Computers in Human Behavior* 62 (2016): 155–167.

29. For a gripping account of this sad episode, see Deborah Blum, *The Poisoner's Handbook: Murder and the Birth of Forensic Medicine in Jazz Age New York* (New York: Penguin, 2010): 152–175.

30. Verne C. Kopytoff, "Sites Like Twitter Absent from Free Speech Pact," *New York Times*, March 6, 2011, http://www.nytimes.com/2011/03/07/technology /07rights.html?_r=1.

31. Stephen Hawkins, Daniel Yudkin, Míriam Juan-Torres, and Tim Dixon, *Hidden Tribes: A Study of America's Polarized Landscape* (New York: More in Common, 2018), https://hiddentribes.us/.

32. To be specific, Mel Epstein initiated this nightly ten o'clock announcement as director of on-air promotions at New York's WNEW-TV.

33. Instagram's own explanation of the change was that it was conducting an experiment (the change was introduced in just a subset of countries). Instagram, "Giving People More Control on Instagram and Facebook," Instagram (Meta) news release, May 26, 2021, https://about.fb.com/news/2021/05/giving-people -more-control/.

34. Shannon Liao, "Twitter Is Thinking about Killing the Like Button—but Don't Hold Your Breath," October 29, 2018, https://www.theverge.com/2018 /10/29/18037458/twitter-like-button-jack-dorsey.

35. Ross and Nisbett, *The Person and the Situation*, 18.

36. Robert R. Johnson, *User-Centered Technology* (Albany, NY: SUNY Press, 1998): 89.

Chapter 8

1. Gary Shteyngart, "O.K., Glass," *New Yorker*, July 29, 2013, https://www .newyorker.com/magazine/2013/08/05/o-k-glass.

2. PA News, "Repost Hateful Messages about Disorder Online and You Could End Up in Court," *Inverness (Scotland) Courier*, August 7, 2024, https://www .inverness-courier.co.uk/news/national/repost-hateful-messages-about-disorder -online-and-you-could-end-up-in-court-114431/.

3. Hugh G. J. Aitken, *The Continuous Wave: Technology and American Radio, 1900–1932* (Princeton, NJ: Princeton University Press, 1985).

4. Friedrich Nietzsche, *Sämtliche Werke: Kritische Studienausgabe in 15 Bänden*, edited by Giorgio Colli and Mazzino Montinari. (Berlin / Munich: de Gruyter, 1988, 89).

5. For current information, consult *Emojipedia*, "FAQs," https://emojipedia .org/faq. See also a post by Emojipedia's founder on the deep history of the emoji. Jeremy Burge, "Correcting the Record on the First Emoji Set," Emojipedia, March 8, 2019, https://blog.emojipedia.org/correcting-the-record-on-the-first -emoji-set/.

6. Ann-Derrick Gaillot and Elena Tarasova, "The Top Emojis of 2023," *Meltwater* (blog), December 15, 2023, https://www.meltwater.com/en/blog/top -emojis-2023.

7. Laura Stampler, "New, Totally Inevitable Social Network Will Let Users Communicate Only via Emoji," *Time*, July 1, 2014, https://time.com/2941701 /new-totally-inevitable-social-network-will-let-users-communicate-only-via -emoji/.

8. Aleina Wachtel, "Feature Deep Dive: React to Comments in Word for Windows," *Microsoft 365 Insider* (blog), June 5, 2023, https://insider .microsoft365.com/en-us/blog/react-to-comments-in-word-for-windows.

9. The vendor of the kiosks Aena uses is named, appropriately, HappyOrNot and is based in Finland. See HappyOrNot, "World's Largest Airport Operator Makes 50 Million Passengers Happier," https://www.happy-or-not.com/en /insights/customer-story/passenger-satisfaction-at-worlds-largest-airport -operator/.

10. Ed Koch, *Citizen Koch: An Autobiography* (New York: St. Martin's Press, 1992).

11. See The White House, "2024 State of the Union," 2024, https://www .whitehouse.gov/state-of-the-union-2024/.

12. Joseph Delpreto, "Controlling Drones and Other Robots with Gestures," Massachusetts Institute of Technology Computer Science & Artificial Intelligence Laboratory, August 12, 2020, https://www.csail.mit.edu/research /controlling-drones-and-other-robots-gestures.

13. Conor Hale, "Precision Neuroscience Deploys 4,096 Electrodes in Brain-Computer Interface Procedure," *Fierce Biotech*, May 31, 2024, https://www .fiercebiotech.com/medtech/precision-neuroscience-deploys-4096-electrodes -brain-computer-interface-procedure.

14. Nicholas Weiler, "Personalized Brain Stimulation Alleviates Severe Depression Symptoms," University of California, San Francisco, Weill Institute for Neurosciences, January 18, 2021, https://psych.ucsf.edu/news/personalized -brain-stimulation-alleviates-severe-depression-symptoms.

15. Rosalind W. Picard, *Affective Computing* (Cambridge, MA: MIT Press, 1997).

16. George Orwell, *1984* (Glasgow, Scotland: William Collins, 2021), 8.

17. Britny Kutuchief, Liz Stanton, and Karolina Mikolajczyk, "The 16 Most Important Social Media Trends for 2024," *Strategy* (blog), Hootsuite, November 14, 2023, https://blog.hootsuite.com/social-media-trends/#5_Shares_will _matter_more_than_likes_comments_or_followers.

18. John Freeman, quoted in Mitchell Hartman, "Disliking Instagram's 'Like' Button," *Marketplace*, June 21, 2019, https://www.marketplace.org/2019/06/21 /instagram-like-button-bad-mental-health/.

19. In fact, *Minority Report* has already become a reality to some degree in the very realm the film portrayed, in that predictive policing has been implemented in many cities. Police forces use algorithms that crunch data on all kinds of factors, starting with a neighborhood's income and education levels but also considering current conditions such as weather and traffic, to produce uncanny forecasts of where criminal activity will crop up next.

20. Rachel Treisman, "X Now Hides Your 'Likes' from Other Users, Whether You Like It or Not," *NPR*, June 13, 2024, https://www.npr.org/2024/06/13/nx-s1 -5004515/x-likes-hide-users-elon-musk#:~:text=June%2013%2C%20 202412:08,%2C%20but%20others%20can't.

21. Pranav Dixit, "Facebook Is Using AI to Supercharge the Algorithm That Recommends You Videos," *Engadget*, March 6, 2024, https://www.engadget .com/facebook-is-using-ai-to-supercharge-the-algorithm-that-recommends-you -videos-033027002.html.

22. Zoe G. Phillips, "Alicia Keys' Super Bowl Performance Pitch-Corrected in YouTube Recording," *Hollywood Reporter*, February 12, 2024, https://www .hollywoodreporter.com/news/music-news/alicia-keys-super-bowl-performance -pitch-corrected-youtube-1235823205/.

23. Laura Llach, "Meet the First Spanish AI Model Earning up to €10,000 per Month," *EuroNews*, March 22, 2024, https://www.euronews.com/next /2024/03/22/meet-the-first-spanish-ai-model-earning-up-to-10000-per-month.

24. Daysia Tolentino, "Snapchat Influencer Launches an AI-Powered 'Virtual Girlfriend' to Help 'Cure Loneliness,'" *NBC News*, May 12, 2023, https://www .nbcnews.com/tech/ai-powered-virtual-girlfriend-caryn-marjorie-snapchat -influencer-rcna84180.

25. Eric J. Boothby and Vanessa K. Bohns, "Why a Simple Act of Kindness Is Not as Simple as It Seems: Underestimating the Positive Impact of Our Compliments on Others," *Personality and Social Psychology Bulletin* 47, no. 5 (2020): 826–840, https://doi.org/10.1177/0146167220949003.

Index

Acknowledgments

There's nothing like writing a book about the like button to remind you of the importance of paying sincere, even if brief, compliments. Throughout the process of writing this book, we have been impressed, enlightened, and often amused by many people and the insights they so generously shared.

So here's a big thumbs-up, first, to each of the dozens of smart, experienced people who were kind enough to submit to interviews or otherwise provide us with research input: Brian Arthur, Delia Baldassari, Lynne Biggar, Nealie Bochma, Paul Buchheit, Steve Chen, Nicholas Christakis, Diane Coyle, Ryan Detert, Jeff Dodds, Devain Doolramani, Michelle Drouin, Brooke Erin Duffy, David Edelman, Carolyn Everson, Thomas Fink, Julia Fitzgerald, Susan Fitzpatrick, Amelia Fletcher, Karl Friston, Mirta Galesic, Dan Gardner, Sam Hall, Reid Hoffman, James Hong, Charles Jennings, David Krakauer, Lucy Kueng, Michael Le, Laura Lee, Mathieu Lefèvre, Anna Lembke, Max Levchin, Simon Levin, Phil Libin, Rita McGrath, Melanie Mitchell, Christine Moorman, Ana Muller, Paul Newby, John Oliver, Emma Parrett, Paresh Patel, Jenna Phillips, Josh Plotkin, Benjamin Rapoport, David Ratajczak, Fernando P. Santos, Gary Shteyngart, Russ Simmons, Biz Stone, Jeremy Stoppelman, Bret Taylor, Neil Waller, Tom Ward, and Kevin Warren.

We have nothing but smiling emoji to shower on the team at Harvard Business Review Press, whose support and guidance have been invaluable, starting with a meeting in 2022 with Melinda Merino. Melinda has provided excellent editorial guidance since, along with her colleague Kevin Evers, who gave us a phrase we reflected on often in the process: that our mission in writing about something as ubiquitous and well established as the like button was "to make the familiar strange." Thanks also to Patty Boyd, Stephani Finks, and Cheyenne Paterson from the editorial and design teams, and Julie Devoll, John Shipley, Lindsey Dietrich, and Alexandra Kephart from the marketing team.

We are indebted especially to Julia Kirby for applying her magical storytelling powers to help us tell our winding tale. Thanks also to Charikleia Kaffe, Annelies O'Dea, Grace Goodhall, Maietreyee Malpekar, Talya Stern, and Lex Verb for the energy, time, and expertise they put into helping us research the history of the like button, as well as Shashi Starling, Amanda Wikman, Adam Job, Brigitta Pristyak, Andras Szabadi, Tania Lekhraj, Erica Quinones, and others at the BCG Henderson Institute for supporting us directly and indirectly in many ways.

Finally, many, many taps of the heart icon to our families, who supported our all-absorbing and extended preoccupation with the project.

About the Authors

MARTIN REEVES is chairman of the BCG Henderson Institute, Boston Consulting Group's think tank on business strategy and management. He is also a senior partner in BCG's San Francisco office and has consulted on strategy issues with a range of clients in consumer goods, industrial goods, pharmaceuticals, and financial services in the United States, Japan, and Europe. His research topics include strategies of resilience, corporate vitality and longevity, technology and competitive advantage, corporate statesmanship, and business imagination. He is a three-time TED speaker, author of *Your Strategy Needs a Strategy* and *The Imagination Machine*, and is widely published in *Harvard Business Review, MIT Sloan Management Review*, and *Fortune*, among others. You can reach him via email at bhi@bcg.com or on X/Twitter @MartinKReeves.

BOB GOODSON is president and founder of Quid, a Silicon Valley–based company whose AI models are used by a third of the *Fortune* 50. Before starting Quid, he was the first employee at Yelp, where he played a role in the genesis of the like button and observed firsthand the rise of the social media industry. After Quid received an award in 2016 from the World Economic Forum for "contributions to the future of the Internet," Bob served a two-year term on WEF's Global Future Council for Artificial Intelligence and Robotics. While at Oxford University doing graduate research in language theory, Bob cofounded Oxford Entrepreneurs to connect scientists with business-minded students. You can reach him at info@bobgoodson.com.